NURSING LEADERSHIP AND MANAGEMENT IN ACTION

Patricia A. Jamerson, RNC, MSN

Cynthia A. Hornberger, MS, RN, MBA

Eleanor J. Sullivan, RN, PhD, FAAN

▲▼▲ ADDISON-WESLEY

An imprint of Addison Wesley Longman, Inc.
Menlo Park, California • Reading, Massachusetts • Harlow, England • New York
Don Mills, Ontario • Sydney • Mexico City • Madrid • Amsterdam

Editorial Assistant: Kim Crowder

Production Editor: David Novak

Copy Editor: Megan Rundel

Proofreader: Martha Ghent

Composition: The Left Coast Group, Inc.

Cover Designer: Vargas/Williams/Design

Copyright © 1997 by Addison Wesley Longman

All rights reserved. No part of this publication may be reproduced, stored in a retrieval system, or transmitted, in any form or by any means, electronic, mechanical, photocopying, recording, or any other media or embodiments now known or hereafter to become known, without the prior written permission of the publisher. Manufactured in the United States of America. Published simultaneously in Canada.

Care has been taken to confirm the accuracy of the information in this book. The authors, editors, and publisher, however, cannot accept any responsibility for errors or omissions or for consequences from the application of the information in this book and make no warranty, express or implied, with respect to its contents.

ISBN 0-8053-7231-8

1 2 3 4 5 6 7 8 9 10—VG—00 99 98 97 96

Addison Wesley Longman, Inc.
2725 Sand Hill Road
Menlo Park, California 94025

PREFACE

Nursing Leadership and Management in Action by Patricia A. Jamerson, Cynthia A. Hornberger, and Eleanor J. Sullivan, is a companion text/workbook for students of nursing management and leadership. A learning tool designed to enhance the understanding of leadership and management principles, this text/workbook facilitates the development of the skills needed in today's health care environment. The practical application exercises provide ample opportunity to apply the theoretical concepts of leadership and management in a safe environment. The exercises in this text provide the practice needed to acquire essential leadership and management skills.

Nursing Leadership and Management in Action is composed of eight sections. Each section includes modules which feature a number of student-friendly exercises complete with simple instructions. The exercises and the modules in this interactive text reflect current health care realities. Modules include information on organizational structure, budgeting, staffing, quality management, computers, critical thinking, communication, delegation, stress reduction, and managing change.

Exercises are designed for both group and individual completion; they are appropriate for lab and lecture settings and can be used as individual review and reinforcement exercises as well. Each exercise has an introduction that summarizes the content necessary to perform the accompanying exercise, analysis or discussion questions, and suggested additional readings.

Related Titles

Nursing Leadership and Management in Action is the perfect companion for students using the *Effective Leadership and Management in Nursing*, Fourth Edition by Eleanor J. Sullivan and Phillip J. Decker. Exercises in this companion text/workbook correspond to discussions in the text. However, since these modules are designed as stand-alone materials, they can also be used in conjunction with other leadership and management texts. The exercises will also be useful to students or new managers who want an independent review of leadership and management topics.

Help for Instructors

The instructor's guidelines for the exercises are in the Instructor's Resource Manual for *Effective Leadership and Management in Nursing*, Fourth Edition by Eleanor J. Sullivan and Patricia A. Jamerson. Instructor information in the manual includes activity classification, learning objectives, and instructions for advanced preparation.

Acknowledgments

The following reviewers offered invaluable comments and suggestions which helped us shape the manuscript.

Valerie Corpuz, RN, MSN
Assistant Professor of Nursing
Holy Names College
Oakland, CA

Pamela A. Crocker, RN, MSN
Department of Nursing
Clarkson College
Omaha, NE

Cynthia Q. Woods, RN, PhD
School of Nursing
University of Kansas Medical Center
Kansas City, KS

TABLE OF CONTENTS

SECTION I UNDERSTANDING NURSING MANAGEMENT AND ORGANIZATIONS 1

Module 1 Organizational Structure
- Exercise 1–1 Determining an Organizational Philosophy 3
- Exercise 1–2 Developing a Strategic Plan 5
- Exercise 1–3 Defining Organizational Structures 7
- Exercise 1–4 Comparing Shared Governance with Traditional Organizational Structures 9
- Exercise 1–5 Recognizing the Impact of Organizational Culture 11
- Exercise 1–6 Comparing Nursing Care Delivery Systems 13
- Exercise 1–7 Redesigning Jobs and Systems 14

Module 2 Leading and Managing
- Exercise 2–1 Comparing Leadership Styles and Their Impact 16
- Exercise 2–2 Recognizing Contemporary Leadership Styles 18
- Exercise 2–3 Identifying Leadership Roles 19
- Exercise 2–4 Identifying Management Functions 21
- Exercise 2–5 Contrasting Management Responsibilities in Nursing 23
- Exercise 2–6 Identifying Nurse Manager Functions 25
- Exercise 2–7 Identifying the Impact of Power 26
- Exercise 2–8 Differentiating the Types of Power 28
- Exercise 2–9 Identifying Personal Sources of Power 30

SECTION II USING RESOURCES EFFICIENTLY AND EFFECTIVELY 33

Module 3 Budget Preparation
- Exercise 3–1 Assessing Productivity 35
- Exercise 3–2 Creating a Personnel Budget 38
- Exercise 3–3 Identifying Variance 40

Module 4 Staffing and Scheduling
- Exercise 4–1 Calculating FTEs 42
- Exercise 4–2 Self Governance: Critique of a Staff Schedule 44

Module 5 Improving Quality
- Exercise 5–1 Case Scenario: Application of the CQI Process 46
- Exercise 5–2 Defining Quality 48
- Exercise 5–3 Comparing CQI and Quality Assurance 50

Module 6 Using Computers
- Exercise 6–1 Case Scenario: Letting Technology Work for You 51
- Exercise 6–2 Understanding Computer Terminology 53
- Exercise 6–3 Accessing the World Wide Web 54

SECTION III   THINKING CRITICALLY   57

Module 7   **Decision Making**
- Exercise 7–1   Decision Making Under Uncertainty or Risk   59
- Exercise 7–2   Group Decision Making: The NASA Experience   61

Module 8   **Problem Solving**
- Exercise 8–1   Learning to Solve Problems   63

Module 9   **Critical Thinking**
- Exercise 9–1   Developing Critical Thinking Skills   65
- Exercise 9–2   Using Statistical Tools   67

SECTION IV   DEVELOPING RELATIONSHIPS   69

Module 10   **Communication**
- Exercise 10–1   Evaluating Personal Communication Techniques   71
- Exercise 10–2   The Importance of Effective Organizational Communication   75
- Exercise 10–3   Communicating Effectively Through Writing   76
- Exercise 10–4   Recognizing Communication Patterns   77
- Exercise 10–5   Developing Assertiveness   79

Module 11   **Groups and Teams**
- Exercise 11–1   Differentiating Roles People Assume in Groups   82
- Exercise 11–2   Identifying the Stages of Group Development   86

Module 12   **Managing Conflict**
- Exercise 12–1   Determining When to Intervene in a Conflict   88
- Exercise 12–2   Using Negotiation   91
- Exercise 12–3   Recognizing Role Conflicts   94

SECTION V   MANAGING TIME AND STRESS   95

Module 13   **Setting Priorities**
- Exercise 13–1   Setting Priorities: An In-Basket Exercise   97
- Exercise 13–2   Recognizing Time Wasters   116
- Exercise 13–3   Managing Your Professional Time   119

Module 14   **Delegation**
- Exercise 14–1   Case Scenario: Use of Delegation   121
- Exercise 14–2   Differentiating Responsibility, Accountability, and Authority   124

Module 15   **Stress Assessment and Reduction**
- Exercise 15–1   Reducing Organizational Stress   126

SECTION VI MANAGING HUMAN RESOURCES 129

Module 16 Interviewing
Exercise 16–1 Developing a Structured Interview Guide — 131
Exercise 16–2 Effective Interviewing — 138

Module 17 Evaluations
Exercise 17–1 How to Write a Critical Incident — 140

Module 18 Staff Development
Exercise 18–1 Coaching as a Staff Development Strategy — 142
Exercise 18–2 Providing a Formal Performance Appraisal — 145
Exercise 18–3 Multicultural Issues in Staff Development — 150

Module 19 Motivation
Exercise 19–1 Personal Motivation: Use of Goal Setting — 152
Exercise 19–2 Employee Motivation — 154

SECTION VII IDENTIFYING ETHICAL AND LEGAL ISSUES 157

Module 20 Avoiding Malpractice
Exercise 20–1 Identifying Malpractice — 159
Exercise 20–2 Comparing Standards — 161
Exercise 20–3 Developing Policies and Procedures — 162

Module 21 Minimizing Organizational Risk
Exercise 21–1 Handling Patient and Family Complaints — 163
Exercise 21–2 Understanding Advanced Directives — 165
Exercise 21–3 Understanding the Role of the State Board of Nursing — 169

Module 22 Legalities in Working with Personnel
Exercise 22–1 Reducing Liability in Hiring Decisions — 170
Exercise 22–2 Understanding Union Grievance Hearings — 172
Exercise 22–3 Developing a Grievance Policy and Procedure — 174

SECTION VIII SURVIVING IN A CHANGING HEALTH CARE ENVIRONMENT 175

Module 23 Personal Image and Career Development
Exercise 23–1 Developing a Résumé — 177
Exercise 23–2 Developing a Career Plan — 178

Module 24 Managing Change
Exercise 24–1 Developing Strategies for Change — 180
Exercise 24–2 Managing Resistance — 182

SECTION I: UNDERSTANDING NURSING MANAGEMENT AND ORGANIZATIONS

MODULE 1 ORGANIZATIONAL STRUCTURE

Exercise 1–1 Determining an Organizational Philosophy
Exercise 1–2 Developing a Strategic Plan
Exercise 1–3 Defining Organizational Structures
Exercise 1–4 Comparing Shared Governance with Traditional Organizational Structures
Exercise 1–5 Recognizing the Impact of Organizational Culture
Exercise 1–6 Comparing Nursing Care Delivery Systems
Exercise 1–7 Redesigning Jobs and Systems

MODULE 2 LEADING AND MANAGING

Exercise 2–1 Comparing Leadership Styles and Their Impact
Exercise 2–2 Recognizing Contemporary Leadership Styles
Exercise 2–3 Identifying Leadership Roles
Exercise 2–4 Identifying Management Functions
Exercise 2–5 Contrasting Management Responsibilities in Nursing
Exercise 2–6 Identifying Nurse Manager Functions
Exercise 2–7 Identifying the Impact of Power
Exercise 2–8 Differentiating the Types of Power
Exercise 2–9 Identifying Personal Sources of Power

MODULE 1 — ORGANIZATIONAL STRUCTURE

Exercise 1–1	Determining an Organizational Philosophy
Exercise 1–2	Developing a Strategic Plan
Exercise 1–3	Defining Organizational Structures
Exercise 1–4	Comparing Shared Governance with Traditional Organizational Structures
Exercise 1–5	Recognizing the Impact of Organizational Culture
Exercise 1–6	Comparing Nursing Care Delivery Systems
Exercise 1–7	Redesigning Jobs and Systems

■ Exercise 1–1 Determining an Organizational Philosophy

INTRODUCTION

The purpose for an organization's existence is related in its *mission statement*. The organization's *vision statement* describes its desired future. Together the mission and vision statements define the organization's *philosophy*, or its values and beliefs. From this philosophy stem *organizational goals* (statements specifying how the mission and vision are to be achieved), unit or *departmental objectives*, and *strategies*. This framework is important in defining the organization's structure and in planning and guiding the organization's work.

EXERCISE

TYPE OF ACTIVITY: INDIVIDUAL OR SMALL GROUP

Instructions

1. Identify the role General Hospital should assume in the community for the 21st century (e.g., continue as a full-service hospital, limit services, specialize).
2. Write a mission statement that conveys that role.
3. Write a feasible vision statement for General Hospital.
4. Identify a goal that supports General Hospital's new philosophy.
5. Write an objective for a unit at General Hospital.

Scenario

With the changes occurring in health care, the board of trustees has asked General Hospital to review its present mission and vision statements. You have been assigned to the committee to work on this project. The organization is currently a 60-bed rural hospital that provides health care services for patients of all ages. The nearest urban hospital is 15 minutes away and it is a teaching hospital also serving patients with a variety of needs and ages. The current mission statement for your organization is "General Hospital exists to promote the health and well-being of the people in our community." The vision has been "to provide quality care, with the best technology available." However, with the escalating costs in health care and increased competition, General Hospital has had difficulty surviving, let alone addressing its vision.

Analysis/Discussion

1. How does the philosophy impact the goals and objectives of the organization?
2. Did the philosophy help or hinder goal development? How?
3. Are your goals and objectives specific and measurable?

Further Reading

El-Namki, M. S. S. (1992). Create a corporate vision. *Long Range Planning,* 25, 25–29.

Martin, L. & Hughes, S. (1993). Using the mission statement to craft a least-restraint policy. *Nursing Management,* 24(3), 65–66.

Matejka, K., Kent, L. & Gregory, B. (1993). Mission impossible? Designing a great mission statement to ignite your plans. *Management Decision,* 31, 34–37.

Peters, J. (1993). On vision and values. *Management Decision,* 31, 14–17.

Exercise 1-2 Developing a Strategic Plan

INTRODUCTION

Strategic planning is a proactive, systematic process whereby management in an organization defines and prioritizes long-term goals of the organization and develops strategies for their implementation. Steps in strategic planning involve identifying the impetus for the strategic plan and relating the plan to the organization's philosophy; assessing the internal and external environments to identify strengths, demands, resources and sources of competition; identifying goals, objectives, and expected outcomes; defining program requirements; and developing implementation, financial and marketing plans.

EXERCISE

TYPE OF ACTIVITY: SMALL OR LARGE GROUP

Instructions

1. Identify the strengths, demands, resources, and sources of competition for the health department.

2. Develop an implementation plan. Identify the steps to take along with who is responsible for each action and any costs involved. Do not forget to identify policies and procedures that may be needed.

Scenario

Nursing administration at a county health department would like to provide the public health nurses with cellular phones because the nurses often have to travel miles to find a phone in an emergency. To provide the nurses with phones will involve strategic planning. The mission of the organization is to provide health care for those in the community who could not otherwise afford it. The vision of the organization is to provide up-to-date, timely services to those served. The county is a 300-square-mile rural county of 2500 people. Many people in the county do not have phones. There is no hospital in the county and only one general practitioner. The four public health nurses who cover the county have each worked at the health department for a minimum of ten years. The nurses know the geography and client population well. The state health department is aware of the county's needs and is willing to provide the finances if a suitable plan is submitted.

Discussion

1. How does getting the public health nurses cellular phones relate to the organization's philosophy?
2. What is the expected outcome of providing cellular phones to the public health nurses?
3. How can the outcome be evaluated?

Further Reading

Curtin, L. (1994). Learning from the future. *Nursing Management,* 25(1), 7–9.

Smith, H. L., Mabon, S. A. & Piland, S. F. (1993). Nursing department strategy, planning, and performance in rural hospitals. *Journal of Nursing Administration,* 23, 23–34.

Thomas, A. M. (1993). Strategic planning: A practical approach. *Nursing Management,* 24, 34–38.

Exercise 1-3 Defining Organizational Structures

INTRODUCTION

When a group of people work together under a defined structure to achieve specific outcomes, that group is an *organization*. The way the group is organized is referred to as its *organizational structure* and that structure is depicted by an *organizational chart*. The structure of an organization reflects its mission, vision, and values, or in other words its philosophy. The structure is also a way of organizing resources to achieve predetermined outcomes. Commonly used structures in health care organizations are functional, service-integrated, hybrid, matrix, and parallel. These structures are classified by the complexity of relationships, the formalization of rules, and the centralization of decision making, authority, and responsibility. The *chain of command* is the established pathway for power, authority, responsibility, and communication. The *span of control* identifies the number of employees for which one manager is responsible.

EXERCISE

TYPE OF ACTIVITY: INDIVIDUAL

Instructions

1. Set up an interview with a manager in a health care organization.
2. During the interview, identify the following:
    a. Type of health care organization (e.g., non-profit hospital, HMO)
    b. Chain of command
    c. Type of decision making
    d. Division of labor
    e. Manager's span of control
    f. Organization's mission
    g. Organization's vision
    h. Organization's strengths and weaknesses due to its structure

Analysis

1. From the information gathered identify the type of departmentalization used (functional, matrix, etc.) Give rationales for your decision.
2. How does this type of organizational structure affect communication? Decision-making?

Further Reading

Hattrup, G. P. & Kleiner, B. H. (1993). How to establish the proper span of control for managers. *Industrial Management, 35,* 28–29.

Jaques, E. (1990). In praise of hierarchy. *Harvard Business Review, 68,* 127–133.

Rakich, J. S., Longest, B. B. & Darr, K. (1992). *Managing Health Services Organizations* (3rd ed.). Baltimore, MD: Health Professions Press.

Exercise 1-4 Comparing Shared Governance with Traditional Organizational Structures

INTRODUCTION

Shared governance is a decentralized organizational structure in which power and authority for clinical practice are transferred to the staff. It is based on the belief that staff are members of a team, and all are valued for the contributions they offer. In addition, it is believed that professional practice can be enhanced through staff decision making. The organizational structure is defined by the staff rather than the hierarchy. Managers and staff collaboratively work together strengthening the links between the organization and staff. Equity, accountability, ownership, and the partnership are the driving factors in improving quality, enhancing interdisciplinary work, empowering staff, and building supportive structures.

EXERCISE

TYPE OF ACTIVITY: SMALL OR LARGE GROUP

Instructions

1. Using the organizational approach assigned to your group, identify committees or individuals needed to address this problem.
2. Use the appropriate committees or individuals to develop a solution to the problem.
3. Be prepared to address each of the following discussion questions.

Scenario

St. Mary's Hospital has always had an excellent reputation in the community. However, with the shortened lengths of stay dictated by third party reimbursement, infection rates and return admissions have skyrocketed. What should the hospital do?

Discussion

1. How did each group address the problem? How long did each group take to reach a decision?
2. Who was involved in the problem-solving process?
3. Who made the ultimate decision?
4. What was the decision?
5. What percent of the group was satisfied with the decision?
6. Do the types of employees in an organization affect the types of structures possible?

Further Reading

DeBaca, V., Jones, K. & Tornebini, J. (1993). A cost-benefit analysis of shared governance. *Journal of Nursing Administration,* 23 (7/8), 50–57.

Jones, C. G., Stasiowski, S., Simons, B. J., Boyd, N. J. & Lucas, M. D. (1993). Shared governance and the nursing practice environment. *Nursing Economics,* 11(4), 208–214.

Porter-O'Grady, T. & Tornebini, J. (1993). Outcomes of shared governance: Impact on the organization. *Seminars for Nurse Managers,* 1(2), 63–73.

Exercise 1–5 Recognizing the Impact of Organizational Culture

INTRODUCTION

The way individuals in a work group think, believe, and behave is known as the *organizational culture*. The culture is a reflection of the employees' shared values and the organization's rituals, policies, and informal rules. Positive or constructive cultures focus on self-actualization, humanism, affiliation, and achievement. Passive-defensive cultures focus on approval, dependency, convention, and avoidance. Aggressive-defensive cultures focus on competition, perfectionism, power, and opposition. The *organizational climate* differs from the organizational culture in that it is a reflection of the employees' perception of an organization. Managers play an important role in dispelling defensive cultures, enhancing a positive climate, and helping new employees understand the organizational culture.

EXERCISE

Type of Activity: Individual or Small Group

Instructions

Answer these questions for your leadership or management clinical setting.[1]

   a. Is the focus of care physical or psychosocial?
   b. How important is organization and efficiency?
   c. How important is it to understand and respect the client's point of view and values?
   d. What is the manager's leadership style?
   e. Who has power? Why?
   f. Who decides what type of nursing care a patient needs?
   g. Do staff work alone or together?
   h. How acceptable is competition?
   i. Do staff criticize each other? Is it open and direct? indirect? in private?
   j. How and by whom is approval given?
   k. Do most of the nurses value past ways of doing things, or do they encourage change?
   l. How important are policies and procedures?
   m. To what extent do staff feel there is one right way to do things?

Analysis/Discussion

1. Based on your responses, how would you define the organizational culture in this setting?
2. What, if anything, could be done to make the culture more positive?
3. What is the organizational climate like?

Further Reading

[1]Coeling, H. V. (1990). Organizational culture: Helping new graduates adjust. *Nurse Educator,* 15 (2), 27.

McDaniel, C. & Stumpf, L. (1993). The organizational culture. *Journal of Nursing Administration,* 23(4), 54–60.

Thomas, C., Ward, M., Chorba, C. & Kumiega, A. (1990). Measuring and interpreting organizational culture. *Journal of Nursing Administration,* 20(6), 17–24.

Exercise 1–6 Comparing Nursing Care Delivery Systems

INTRODUCTION

Nursing care is delivered amidst a number of different structural designs in health care. Each design has its own strengths and weaknesses. In *functional nursing* different health care givers are assigned tasks according to their skills and licensure. In *team/modular nursing* groups of health care providers are assigned to a group of patients. The *case* or *total patient care method* makes one nurse completely responsible for all aspects of nursing care required by one or more patients. In *primary nursing* one nurse coordinates all aspects of a patient's care and develops a plan for others to follow. *Practice partnerships* between a RN and an assistant are designed to achieve the same outcomes of two RNs by selectively delegating less specialized tasks to the assistant under the direction of the RN. *Differentiated practice* also capitalizes on the different educational, experiential, and competence levels of health care providers by defining their roles and functions accordingly. *Case management* is a method of coordinating and monitoring patient services to achieve desired outcomes, within a specific time frame, that are usually defined in critical pathways.

EXERCISE

TYPE OF ACTIVITY: INDIVIDUAL

Instructions

Pretend you have been hired as the new nurse manager for a pediatric rehabilitation unit in a long-term care facility. Select the nursing care delivery system you would prefer to implement.

Analysis

1. Provide a rationale for your decision.
2. What are the strengths and weaknesses associated with this nursing care delivery system?

Further Reading

Abts, D., Hofer, M., & Leafgreen, P. K. (1994). Redefining care delivery: A modular system. *Nursing Management,* 25, 40–46.

Huber, D. G., Blegen, M. A. & McCloskey, J. C. (1994). Use of nursing assistants: Staff nurse opinions. *Nursing Management,* 25, 64–68.

Petryshen, P. R. & Petryshen, P. M. (1992). The case management model: An innovative approach to the delivery of patient care. *Journal of Nursing Administration,* 17(10), 1188–1194.

Pitts-Wilhelm, N. C. & Koerner, J. (1991). Differentiating nursing practice to improve service outcomes. *Nursing Management,* 22(12), 22–25.

Exercise 1–7 Redesigning Jobs and Systems

INTRODUCTION

In an effort to reduce health care costs, organizations are evaluating the overall effectiveness and efficiency of their structures and work groups. In *job redesign,* efforts focus on more efficient utilization of staff. The degree of specialization, distribution of activities, role overlap, and waste are examined. *System redesign* addresses changes needed in the actual structure of the organization (organizational restructuring) or its processes (system reengineering).

EXERCISE

TYPE OF ACTIVITY: SMALL OR LARGE GROUP

Instructions

1. Divide into groups as directed.
2. Read the following scenario.
3. Develop a plan, according to your group's assignment, for redesign.
4. Identify the benefits and barriers associated with your plan.
5. Identify a spokesperson who will relay your group's plan.

Scenario

General Hospital needs to change their current operating procedures to survive fiscally. They are unsure how to proceed. Three task forces are established to decide whether to redesign jobs, restructure, or reengineer. Currently, General is a non-profit, 200-bed community hospital with a highly bureaucratic, parallel organizational structure (Figure 1). Modular nursing is used.

Discussion

1. What are the benefits of each plan? the barriers?
2. How will each of the proposed changes affect the organization and its employees?
3. How will the proposed changes improve the organization's financial status?

FIGURE 1 Organizational Structure for General Hospital

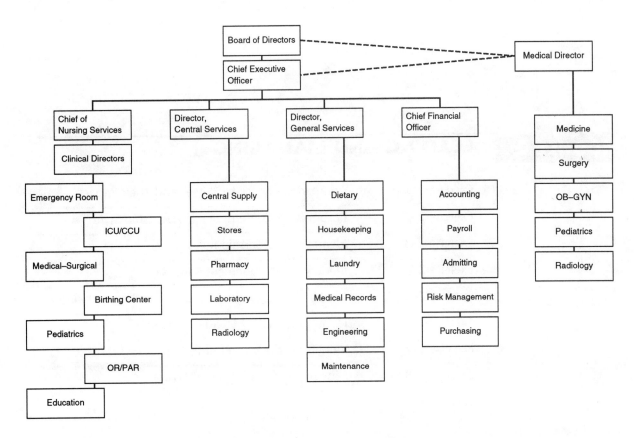

Further Reading

Greenberg, L. (1994). Work redesign: An overview. *Journal of Emergency Nursing,* 20(3), 28a–32a.

Hammer, M. & Champy, J. (1993). *Reengineering the Corporation: A Manifesto for Business Revolution.* New York: Harper Collins.

MODULE 2 — LEADING AND MANAGING

Exercise 2–1	Comparing Leadership Styles and Their Impact
Exercise 2–2	Recognizing Contemporary Leadership Styles
Exercise 2–3	Identifying Leadership Roles
Exercise 2–4	Identifying Management Functions
Exercise 2–5	Contrasting Management Responsibilities in Nursing
Exercise 2–6	Identifying Nurse Manager Functions
Exercise 2–7	Identifying the Impact of Power
Exercise 2–8	Differentiating the Types of Power
Exercise 2–9	Identifying Personal Sources of Power

■ Exercise 2–1 Comparing Leadership Styles and Their Impact

INTRODUCTION

Behavioral scientists studying management have identified four different styles of leadership: autocratic, bureaucratic, democratic, and laissez-faire. Each is based on the manager's beliefs about employee motivation and skills. The *autocratic* leader believes employees are externally motivated so the leader maintains strong control, gives directions and uses coercion. The *bureaucrat* also believes employees are externally motivated, but trusts neither followers nor self to make decisions and therefore relies on policies and organizational rules to direct work. Both *democratic* and *laissez-faire* managers believe workers are internally motivated so they involve employees in decision making.

However, the laissez-faire manager provides no direction, control, or criticism. The democratic leader provides suggestions, guidance, and constructive criticism.

EXERCISE

TYPE OF ACTIVITY: SMALL OR LARGE GROUP

Instructions

1. Get into groups as assigned. The instructor will ask for a volunteer "leader" from each group.
2. The leader will be given further instructions on role-playing the assigned leadership style from the instructor.
3. The instructor will describe a project for each group to complete.
4. The leader will direct the group's work, role-playing the assigned leadership style.

Discussion

At the end of the designated time period,
1. Each group should identify its leader's leadership style.
2. The groups should compare their motivation, productivity, job satisfaction, and trust in the leader. What were the similarities and differences?

Further Reading

Tannebaum, R. & Schmidt, W. (1973). How to choose a leadership pattern. *Harvard Business Review*, 51, 16.

Exercise 2–2 Recognizing Contemporary Leadership Styles

INTRODUCTION

Today, leadership is viewed as a combination of trait, behavior, and contingency theories. *Charismatic leadership* is based on the leader's personal qualities that arouse affection for and commitment to the leader and her or his beliefs. *Transactional leadership* is based on social exchange theory. Successful leaders understand followers' needs and use incentives to enhance employee loyalty and performance and maintain the status quo. *Transformational leaders* inspire their followers to perform beyond basic expectations and to commit to the organizational vision. Dependency is discouraged and the followers are encouraged to exercise leadership. *Connective leaders* use their interpersonal skills to create connections, encourage collaboration, and foster integration. In *shared leadership,* the leadership potential of professionals is recognized and used in a variety of situations, such as self-directed work teams and shared governance. *Servant leadership* is based on the premise that while serving, one may be called to lead and effect change in other individuals, systems, and organizations.

EXERCISE

Type of Activity: Individual

Instructions

1. Identify an individual you consider a leader in an organization (e.g., school, hospital, community group).
2. Identify which contemporary leadership style(s) (transactional, transformational, connective, shared, or servant) fit the leader. Support your answer with examples.

Analysis

Are the style(s) used effective within this organization? Why or why not?

Further Reading

Davidhizar, R. (1993). Leading with charisma. *Journal of Advanced Nursing,* 18, 675–679.

Kiechel, W. (1992). The leader as servant. *Fortune,* 125, 121–122.

Marriner-Tomey, A. (1993). *Transformational Leadership in Nursing.* St. Louis: Mosby-Year Book.

Exercise 2–3 Identifying Leadership Roles

INTRODUCTION

A leader is someone who uses interpersonal skills to influence others to accomplish a specific goal. Leadership is a dynamic process in which a variety of personal behaviors and strategies are used. Common skills used are communicating, motivating, initiating, facilitating, and integrating.

EXERCISE

TYPE OF ACTIVITY: INDIVIDUAL

Instructions

1. Think of someone you consider to be a leader.
2. Using Table 1, check each of the following roles and characteristics that apply to this leader.
3. Check which of the following characteristics apply to you.

TABLE 1   LEADERSHIP ROLES

CHARACTERISTICS	LEADER	SELF	CHARACTERISTICS	LEADER	SELF
■ Able to enlist cooperation			■ Good interpersonal skills		
■ Advocate			■ Good judgment		
■ Buffer			■ Independent		
■ Change agent			■ Knowledgeable		
■ Coach			■ Mentor		
■ Communicator			■ Motivator		
■ Cooperative			■ Negotiator		
■ Creative			■ Nonconformist		
■ Critical thinker			■ Problem solver		
■ Decision maker			■ Risk taker		
■ Diplomat			■ Role model		
■ Empowerer			■ Self-confident		
■ Evaluator			■ Socially active		
■ Facilitator			■ Tactful		
■ Flexible			■ Teacher		
■ Forecaster			■ Trustworthy		
■ Good emotional control			■ Visionary		

Analysis

1. How well does the profile of leadership characteristics fit the identified leader?
2. How well does the profile fit you?
3. What can you do to enhance your leadership potential?

Further Reading

Bates, S. & Fosbinder, D. (1994). Using an interview guide to identify effective nurse managers. *Journal of Nursing Administration,* 24(45), 33–38.

Capowski, G. (1994). Anatomy of a leader: Where are the leaders of tomorrow? *Management Review,* 83, 10–17.

Huston, C. J. (1990). What makes the difference? Attributes of the exceptional nurse. *Nursing* 90, 20(5).

Exercise 2–4 Identifying Management Functions

INTRODUCTION

The management functions of planning, organizing, directing, and controlling were first described by Henri Fayol in 1916. *Planning* is the ongoing process of assessing and analyzing the present situation, predicting future trends and events, making decisions and solving problems, establishing goals and objectives, implementing plans, and evaluating outcomes. *Organizing* is the process of coordinating the work to be done and reviewing the use of human and material resources. *Directing* is the process of getting the organization's work done efficiently and effectively. *Controlling* is the regulation of organizational activities by establishing standards, determining the methods to be used in measurement, evaluating results, and providing feedback.

EXERCISE

Type of Activity: Individual

Instructions

Using Fayol's management functions, categorize each of the following components of a manager's job as it relates to planning (P), organizing (O), directing (D), and controlling (C).

____ Developing a budget

____ Interviewing job candidates

____ Conducting performance evaluations

____ Establishing work schedules

____ Improving productivity

____ Initiating change

____ Conducting a staff meeting

____ Managing conflict

____ Setting priorities

____ Improving quality

____ Developing a strategic plan

____ Delegating tasks

____ Setting a meeting agenda

____ Coaching staff

____ Disciplining an employee

____ Establishing policies and procedures

____ Designing job descriptions

____ Maintaining employee files

____ Attending meetings

____ Staffing a unit

Analysis

1. How easy is it to categorize these functions?
2. What kinds of difficulties did you have in categorizing these functions?

Further Reading

Ameduri, P. (1994). Directing others is a demanding role. *RN,* 57(10), 21–24.

Corpaz, L. & Conforti, C. (1994). Organizing and documenting clinical standards. *Nursing Management,* 25(5), 70–72, 74–76.

Exercise 2–5 Contrasting Management Responsibilities in Nursing

INTRODUCTION

Traditionally, there have been three levels of management: upper level, middle level, and first level. The *upper-level manager* is not only responsible for establishing organizational goals and strategic plans for nursing, but for buffering the effects of the external environment. *Middle-level managers* supervise first-level managers and serve as a liaison between upper- and first-level managers. *First-level managers* are responsible for supervising the work of nonmanagerial personnel and the day-to-day activities of a specific work unit or units. Today in an atmosphere of downsizing and restructuring, the levels of management are being redefined. Staff nurses are taking on more management functions. Charge nurses are being considered management, and middle management positions are being abolished.

EXERCISE

TYPE OF ACTIVITY: SMALL OR LARGE GROUP

Instructions

Ask the managers on the panel the following questions:
a. How are you accountable for the clinical practice of nursing and the delivery of patient care?
b. How are you accountable for managing human, fiscal, and other resources needed to manage clinical nursing practice and patient care?
c. How do you facilitate the development of licensed and unlicensed personnel?
d. How do you ensure institutional compliance with professional, regulatory, and government standards of care?
e. What and how do you contribute to strategic planning?
f. How do you facilitate cooperative and collaborative relationships among disciplines and departments?

Discussion

1. How are the roles and functions of nurse managers similar across the levels?
2. How do nurse manager roles at different levels differ?

Further Reading

Beaman, A. L. (1986). What do first-line managers do? *Journal of Nursing Administration,* 16, 6–9.

Mark, B. A. (1994). The emerging role of the nurse manager. *Journal of Nursing Administration,* 24(1), 48–55.

Exercise 2–6 Identifying Nurse Manager Functions

INTRODUCTION

Nurse managers have 24-hour accountability for the management of specific work groups within a health care organization. The role is multifaceted. The responsibilities of nurse managers include (a) ensuring excellence in the clinical practice and delivery of patient care in the area(s) of responsibility; (b) managing resources; (c) facilitating personnel development; (d) ensuring compliance with professional, regulatory, and government standards; (e) developing and implementing strategic plans; and (f) facilitating cooperative and collaborative relationships.[1] To fulfill these responsibilities involves juggling a number of roles and functions. The manager must constantly assess, make decisions, solve problems, plan, organize, facilitate, and motivate. In addition, the manager must manage conflict, resources, and time.

EXERCISE

TYPE OF ACTIVITY: INDIVIDUAL OR SMALL GROUP

Instructions

1. Follow a nurse manager for a shift.
2. Document the types of activities in which the manager engages.
3. Prioritize each of the following, on a scale of 1 (most) to 10 (least), based on the amount of time devoted to each function.

   ___ Setting goals   ___ Managing financial resources
   ___ Making decisions   ___ Managing her/his time
   ___ Managing conflict   ___ Motivating
   ___ Scheduling staff   ___ Developing staff
   ___ Communicating   ___ Planning

Analysis/Discussion

1. What did you learn about the nurse manager's role?
2. How was the majority of the manager's time spent?
3. Was your observation consistent with what you previously thought a nurse manager did? Why or why not?

Further Reading

[1]American Organization of Nurse Executives. (1992). The roles and functions of the hospital nurse manager. *American Hospital Association Advisory.* Chicago: AHA.

Dieneman, J. & Schaffer, C. (1992). Manager responsibilities in community agencies and hospitals. *Journal of Nursing Administration,* 22(5), 40–45.

Exercise 2–7 Identifying the Impact of Power

INTRODUCTION

Power is the potential to achieve goals, the ability to influence others to perform a task they otherwise may not. Power is often used to gain a competitive advantage, acquire information, motivate, communicate, and improve performance or processes. However, power is often misused and as a result may be feared or mistrusted. Recognizing the impact of power plays is necessary to obtain the desired outcomes. Persuasion should be used over coercion, patience over impatience, openness over close-mindedness, compassion over confrontation, and integrity over dishonesty.

EXERCISE

TYPE OF ACTIVITY: INDIVIDUAL OR SMALL GROUP

Instructions

1. Using Table 1, identify the power plays in each of the following scenarios.
2. Describe the possible recipient outcomes.

TABLE 1   POWER PLAYS

Power Play	Recipient's Response
▪ "Let's be fair"	▪ Feelings of insecurity; insecure about choices because power game is played by someone else's rule.
▪ "Can you prove that?"	▪ Embarrassed by inability to defend self.
▪ "Be specific"	▪ Feelings of incompetence if facts and figures cannot be generated to support position.
▪ "It's either this or that . . . which is it? Take your pick"	▪ Angered at being forced to pick between limited options.
▪ "But you said . . . and now you say . . ."	▪ Confused about what was meant; believe position is illogical.

Scenarios

1. Ann learned in a recent staff meeting that taxicab tickets would no longer be provided by social services. However, it would be a while before social services could provide a written copy of the policy or change their brochure listing their services. The very next day, Mr. Green asks Ann for a taxicab ticket. She tries to explain the change in policy. Mr. Green retorts he has received taxicab tickets before. He wants to see a copy of the policy. When Ann is unable to produce the policy or reach anyone from social services, Mr. Green claims Ann is infringing upon his rights.

2. In a chance meeting with the chief of surgery, Maria relates stories told to her by her nurses about some of the residents. Rather than expressing interest, the chief of surgery becomes defensive and wants to know when the situations occurred, who was involved, and if Maria could not supply those details, he didn't want to hear anymore about it.

3. Susan questions Nathan, her manager, about why she is scheduled every Saturday during the month of December, when everyone else is working every other weekend. Nathan retorts, "It's every Saturday or Christmas—what's your pleasure?"

4. Sabrina has worked in the intensive care unit for 11 months and is looking forward to orienting to charge, but her manager insists that before orienting, a nurse must work in her intensive care unit for a minimum of one year. When Sabrina arrived at work today, she was surprised to find Paul orienting to charge. Although Paul has previous ICU experience, he has only worked in the unit 10½ months. Sabrina confronted her manager about the apparent change in rules.

Analysis/Discussion

1. How easy or difficult was it to identify the power plays in these scenarios?
2. Have you ever been a recipient of one of these power plays? How did you react?

Further Reading

Farmer, B. (1993). The use and abuse of power in nursing. *Nursing Standard, 7,* 33–36.

Palich, L. E. & Hom, P. W. (1992). The impact of leader power and behavior on leadership. *Group and Organization Management, 17,* 279–296.

Sneed, N. V. (1991). Power: Its use and potential for misuse by nurse consultants. *Clinical Nurse Specialist, 5,* 58–62.

Exercise 2–8 Differentiating the Types of Power

INTRODUCTION

Nurses, wherever they practice, use power in all that they do. There are seven types of power frequently exhibited: coercive, connection, expert, information, legitimate, referent, and reward. Often, combinations of these power sources are used and there may be some overlap. *Coercive power* is the use of fear, coercion, and punishment to achieve an outcome. Conversely, *reward power* is power associated with an individual's ability to grant rewards to others. *Expert power* is derived from the knowledge and skills one has that are needed by others. Similarly, *information power* is that power that comes from having information that others need. *Legitimate power,* or authority, is the power that accompanies a position within an organization or group. *Referent power* is ascribed by others to individuals seen as charismatic or popular. Associating with others perceived as powerful provides *connection power.*

EXERCISE

Type of Activity: Individual

Instructions

In each of the following scenarios, identify (a) what type of power is being used and (b) what the motivation is to comply.

Scenarios

1. In a recent staff meeting, Sean asks his staff to each identify one night each pay period when they can work to cover staffing.
2. Marita gave those staff members who worked so hard on developing policies and procedures before JCAHO visited, a day off with pay.
3. Stephen sits down with each of his new employees and answers any questions they may have about the unit.
4. Following each advisory committee meeting, Joyce provides a copy of the minutes in the communication book.
5. Mark documents all absences and late arrivals in his employee files.
6. Juana is very involved in a number of civic and professional organizations.
7. Besides being the chief nurse executive, Janelle is president of the state nurse's association.

Analysis

1. How does power increase each individual's influence?
2. How easy or difficult was it to identify the different types of power?

Further Reading

del Bueno, D. J. (1987). How well do you use power? *American Journal of Nursing,* 87(11), 1495–1496.

Exercise 2–9 Identifying Personal Sources of Power

INTRODUCTION

Power is based on honor, respect, loyalty, and commitment. But how does one gain power? Getting to know people, the organization, and yourself is a start. Image is also important. Grooming, dress, manners, body language, and speech are significant characteristics in developing a powerful image. Equally important are interpersonal skills such as communicating and listening. Additional strategies are to increase your visibility through involvement and networking.

EXERCISE

TYPE OF ACTIVITY: INDIVIDUAL

Instructions

Complete the self-evaluation below, reflecting upon your activities as a student and/or employee. Rate yourself for each characteristic on a scale of 1 (poor) to 5 (excellent). Review your assessment and then address the analysis questions.

Image
- ___ Hygiene
- ___ Grooming
- ___ Dress
- ___ Poise
- ___ Manners
- ___ Nonverbal communication skills
- ___ Verbal communication skills
- ___ Listening abilities
- ___ Self-concept
- ___ Attitude
- ___ Motivation
- ___ Knowledge
- ___ Abilities
- ___ Health
- ___ Credibility
- ___ Flexibility

Savvy
- ___ Level of involvement
- ___ Productivity
- ___ Efficiency
- ___ Effectiveness
- ___ Knowledgeable about organization
- ___ Visionary
- ___ Visibility in the organization
- ___ Asks for help and advice
- ___ Associates with powerful individuals
- ___ Networking
- ___ Empowerment of others
- ___ Continuous development of self

Analysis

1. Did you identify any new sources of power during your assessment?
2. What resources do you need before you can expand your power base?
3. Develop a plan for increasing your power base.

Further Reading

Hoelzel, C. B. (1989). Using structural power sources to increase influence. *Journal of Nursing Administration,* 9(11), 10–15.

Skelton, R. (1994). Nursing and empowerment: Concepts and strategies. *Journal of Advances in Nursing,* 19, 415–423.

SECTION II

USING RESOURCES EFFICIENTLY AND EFFECTIVELY

MODULE 3   **BUDGET PREPARATION**

Exercise 3–1   Assessing Productivity
Exercise 3–2   Creating a Personnel Budget
Exercise 3–3   Identifying Variance

MODULE 4   **STAFFING AND SCHEDULING**

Exercise 4–1   Calculating FTEs
Exercise 4–2   Self Governance: Critique of a Staff Schedule

MODULE 5   **IMPROVING QUALITY**

Exercise 5–1   Case Scenario: Application of the CQI Process
Exercise 5–2   Defining Quality
Exercise 5–3   Comparing CQI and Quality Assurance

MODULE 6   **USING COMPUTERS**

Exercise 6–1   Case Scenario: Letting Technology Work for You
Exercise 6–2   Understanding Computer Terminology
Exercise 6–3   Accessing the World Wide Web

MODULE 3 — BUDGET PREPARATION

Exercise 3–1   Assessing Productivity
Exercise 3–2   Creating a Personnel Budget
Exercise 3–3   Identifying Variance

■ Exercise 3–1 Assessing Productivity

INTRODUCTION

In the process of cost containment, nurse managers commonly measure productivity to evaluate nursing services and to provide necessary information to create budgets. These measures provide specific information about a unit's productivity. One such measure is *Nursing Care Hours/Patient Days*. *Nursing Care Hours* are the total hours worked by nursing personnel of a designated unit, including registered nurses, licensed practical nurses, and unlicensed assistive personnel, for a specific time period. The most common time period is one month, but may be calculated on a daily, weekly, or pay-period basis. *Patient Days* is the sum total of days charged per patient within the same time period.

This ratio provides a rough estimate of the intensity of care provided in that time period. It does not, however, tell the nurse manager about what types of care were required. If that information is needed, the nurse manager needs to break down the information for each level of care (for example, a registered nurse) per patient day. Nursing care hours are projected using acuity systems commonly included in health care information systems (HIS). Reports are provided to the nurse manager which include the variance from the budgeted productivity ratio. *Variance* is the difference between actual and budgeted events such as dollars expended or a calculated productivity measure. Monthly variance reports provide ongoing opportunities for nurse managers to identify alterations rapidly so that necessary adaptations can be made.

EXERCISE

TYPE OF ACTIVITY: INDIVIDUAL

Table 1 provides a set of data for a high acuity medical-surgical unit for the previous year. Data included are the number of budgeted and actual patient days, total direct budgeted nursing care hours, and the actual hours of paid nursing care. Included are columns for the calculated productivity measure and the variance per month.

[handwritten note: Act. Nurs. hrs / total # pt. days]

TABLE 1 — PRODUCTIVITY DATA FOR 40-BED MEDICAL-SURGICAL UNIT

Month	Actual Patient Days	Budgeted Patient Days	Actual Nursing Hours	Budgeted Nursing Hours	Productivity Measure	Variance
January	1116	1116	10,965	10,044	9.82	.82 hrs
February	1030	1008	10,209	9074	9.91	.91 hrs
March	974	896	8060	8064	9.2	.2 hrs
April	876	960	8560	8640		
May	930	868	7739	7812		
June	864	840	7553	7560		
July	942	868	7840	7812		
August	831	868	7993	7812		
September	888	840	7868	7560	9.23	.23 hrs
October	967	868	8013	7812		
November	996	960	8642	8640		
December	1091	896	10,238	8064		
TOTAL	11,510	10,988	103,680	98,892	9.13	.43 hrs

Instructions

1. Given the information in Table 1, calculate the total hours of direct nursing care delivered per patient day for each month and for the total year.

2. Assuming that the productivity standard for this unit is 9 hours per patient day, calculate the variance for each month.

Discussion

1. Discuss the trends you see in this data.

2. How can this information be used by the nurse manager to better budget for the upcoming year? What are your suggestions for change?

3. If you were the nurse manager, what other information might you request to better determine productivity on your unit?

Further Reading

Jordan, S. D. (1994). Nursing productivity in rural hospitals. *Nursing Management* 25(3), 58–62.

Townsend, M. B. (1991). Creating a better work environment: Measuring effectiveness. *Journal of Nursing Administration*, 21(1), 11–14.

Exercise 3–2 Creating a Personnel Budget

INTRODUCTION

One of the most challenging aspects of a nurse manager's job is to accurately project the personnel budget for the upcoming fiscal year. Budget preparation requires knowledge of previous and future trends in care delivery, market information to predict changes in services offered, and unit productivity measures. To prepare a personnel budget, the nurse manager is typically provided with historical budget information, productivity measures, and a specified budgetary allowance for provision of services.

EXERCISE

TYPE OF ACTIVITY: INDIVIDUAL

Instructions

Using the information provided in Table 1, project the monthly and total year expenditures for this unit. Be sure to include fringe benefit costs.

Notes

1. Assume that the budgeted nursing care hours are the basis of the projections.
2. The hospital is attempting to reduce length of stay, so patient days are expected to decrease by 2%. Projected budget hours will be 98% of previous budget hours.
3. The RN median wage is $16.76, and takes into account all shift differential and bonus pay. Inflation and raises are expected to raise salaries 2.5% next year.
4. Fringe benefits, including sick, overtime, vacation, insurance, and social security, will add 15% to the total costs.
5. Total hours projected monthly = Projected budget hours × (projected wage + fringe benefits).

Christine Coode

TABLE 1	PERSONNEL BUDGET WORKSHEET				
Month	Previous Budget Hours	Projected Budget Hours	Projected Wage	Fringe Benefits	Total Projected Monthly
▪ January	10,044	9843.12	169,095 +	25,364 =	194,459
▪ February	9072	8890.56	152,731 +	22,910 =	175,641
▪ March	8064	7902.72	135,761 +	20,364 =	156,125
▪ April	8640	8467.20	145,458 +	21,819 =	167,277
▪ May	7812	7655.76	131,518 +	19,728 =	151,246
▪ June	7560	7408.80	127,276 +	19,091 =	146,367
▪ July	7812	7655.76	131,518 +	19,728 =	151,246
▪ August	7812	7655.76	131,518 +	19,728 =	151,246
▪ September	7560	7408.80	127,276 +	19,091 =	146,367
▪ October	7812	7655.76	131,518 +	19,728 =	151,246
▪ November	8640	8467.20	145,458 +	21,819 =	167,277
▪ December	8064	7902.72	135,761 +	20,364 =	156,125
▪ TOTAL	98,892	96,914.16	1,664,888	249,733	1,914,621

handwritten annotations:
19,759
17,179 2,58 = 194,490.20
× .98   16.76   × .15%
× 2.5%
16.76 + 0.419 = 17.179
0.419

98,892
× 16.76
─────
1,657,429.92
× .15%
─────
248,614.488

1,657,429.92
+ 248,614.49
─────
$1,906,044.41

Analysis

1. What is the increase in budget for personnel that you will need to request for this year? 1,914,621 − 1,906,044 = $8,577

2. If the hospital is unable to commit to this amount of increase, what are your options as a nurse manager?

Further Reading

Dreisbach, A. M. (1994). A structured approach to expert financial management: A financial development plan for nurse managers. *Nursing Economics*, 12(3), 131–139.

Willburn, D. (1992). Budget response to volume variability. *Nursing Management*, 43(2), 42–44.

Exercise 3-3 Identifying Variance

INTRODUCTION

For purposes of controlling, variance calculations provide timely information to identify problems and strengths in productivity and efficiency. Variance is the difference between actual and budgeted events such as dollars spent or a calculated productivity measure. Monthly variance reports provide ongoing opportunities for nurse managers to identify alterations rapidly so that necessary adaptations can be made. Variance alterations that indicate a decrease in productivity and efficiency must be justified each month.

EXERCISE

TYPE OF ACTIVITY: INDIVIDUAL

Instructions

Using the information in Table 1, determine the degree and direction of the unit's variance per month as calculated as:

$$\frac{\text{Total payroll expenses for registered nurses in time } X}{\text{Total required hours of care in time } X}$$

1. Calculate the productivity measure.
2. Divide the calculated productivity ratio by the projected productivity ratio. Subtract this value from 1.0 and record the variance, keeping the positive or negative sign of the result.

Discussion

1. What is the actual productivity measure for the year?
2. How well does the overall measure reflect the monthly activity?
3. If you were the nurse manager, what months would you take a closer look at the unit activity?
4. What are some of the reasons that the productivity measure varies from the projected value?
5. As the nurse manager, what are some actions you could take to improve the variance?

Further Reading

Francisco, P. D. (1989). Flexible budgeting and variance analysis. *Nursing Management, 20*(11), 40–43.

TABLE 1   RN SALARY COST PER PATIENT DAY

Month	Total Payroll Expenses	Total Nursing Care Hours	Actual Ratio	Budgeted Ratio	Variance
January	36,234	1786	20.3	19.3	<.05>
February	32,388	1648	19.7	19.3	<.02>
March	36,129	1725	20.9	19.3	<.08>
April	26,987	1402	19.2	19.3	.01
May	28,175	1488	18.9	19.3	.02
June	30,512	1582	19.3	19.3	0
July	29,758	1507	19.7	19.3	<.02>
August	26,981	1328	20.3	19.3	<.03>
September	25,128	1421	17.7	19.3	.08
October	28,790	1546	18.6	19.3	.04
November	30,900	1593	19.4	19.3	0
December	35,384	1747	20.3	19.3	<.05>
TOTAL	367,366	18,773	19.6	19.3	<.02>

MODULE 4 — STAFFING AND SCHEDULING

| Exercise 4–1 | Calculating FTEs |
| Exercise 4–2 | Self Governance: Critique of a Staff Schedule |

■ Exercise 4–1 Calculating FTEs

INTRODUCTION

To develop a master staffing pattern, a nurse manager must determine how many hours of nursing care are required based upon available patient classification system information. Once the hours of required nursing care are determined, the number of full-time equivalents (FTEs) can be calculated. One full-time equivalent is equal to 40 hours per week, 80 hours per 2-week pay period, or 2080 hours per year.

EXERCISE

Type of Activity: Individual

Instructions

1. Based on the information provided in Table 1, calculate the FTEs required for a 2-week schedule.

2. If this unit uses an average of 3 hours per FTE of vacation time and 2 hours per FTE of sick time per pay period, what will be the actual FTEs needed for RNs, LPNs, and unlicensed assistive personnel for the first two weeks of February?

TABLE 1	TOTAL REQUIRED NURSING CARE HOURS BY CLASSIFICATION	
CLASSIFICATION	TOTAL HOURS PER PAY PERIOD	FTEs
▪ RN	704	8.8
▪ LPN	227	2.8
▪ Unlicensed	1134	14.175

Analysis

1. What is the impact of fringe benefits on the FTE requirements for delivery of nursing care?

2. If you had four nurses that wished to work half-time, how many nurses total would you need to hire?

Further Reading

McHugh, M. L. & Dwyer, V. L. (1992). Measurement issues in patient acuity classification for prediction of hours of nursing care. *Nursing Administration Quarterly*, 16(4), 20–31.

$$\frac{704}{80} = 8.8 \text{ FTE}$$

$$\times 5 \text{ HRS}$$

$$44 + 704 = \frac{748}{80} = 9.35 \text{ RN FTEs}$$

$$\frac{227}{80} = 2.8 \text{ FTE}$$

$$\times 5$$

$$14.2 + 227 = \frac{241.19}{80} = 3.01 \text{ FTE}$$

$$\frac{1134}{80} = 14.18$$

$$\times 5$$

$$70.9 + 1134 = \frac{1204.9}{80} = 15 \text{ FTE}$$

Exercise 4–2 Self-Governance: Critique of a Staff Schedule

INTRODUCTION

Staff scheduling is a routine responsibility of the nurse manager. Currently, many managers delegate scheduling to the staff in a self-governance strategy. However, it is the nurse manager's responsibility to evaluate the schedule for accuracy, fairness, and adequate coverage. It is also the manager's responsibility to knowledgeably delegate this responsibility and skillfully provide needed expertise to make this delegation successful. One of the first responsibilities is to become comfortable with evaluation of a proposed schedule.

EXERCISE

TYPE OF ACTIVITY: INDIVIDUAL

Instructions

The staff has created the following schedule for the upcoming 2 weeks. You are responsible for evaluating the schedule prior to posting. As you begin to analyze the schedule given in Table 1, you are aware of the following personnel requirements and personal requests:

1. The unit is staffed with 7.7 FTEs of daytime RN coverage.
2. Carol, Stefan, and Celia have agreed to work every weekend. Everyone else has every other weekend off.
3. The unit is staffed with a combination of 7 AM–3 PM and 7 AM–7 PM shifts. However, only one RN is scheduled for 7 AM–7 PM per day.
4. No staff member can work more than 40 hours per week per federal regulation.
5. Adequate coverage consists of five RNs for the day shift.

TABLE 1 — DAYSHIFT SCHEDULE FOR 6 WEST

1 wk (columns S–S) / *2 wk* (columns S–S)

RN	S	M	T	W	R	F	S	S	M	T	W	R	F	S	
Carol	7–3	7–3	7–3	7–3	7–3	7–3	7–3	V	D	D	D	7–3	7–3	7–3	= 80 HRS
Beth	7–3	D	7–3	D	7–3	7–3	D	D	D	7–3	7–3	7–3	7–3	7–3	= 72 HRS
Steve	7–3	7–3	7–3	D	7–3	7–3	7–3	7–3	D	7–3	7–3	D	D	7–3	= 80 HRS
Karen	D	7–3	7–3	7–3	7–3	D	7–3	7–3	7–3	7–3	7–3	7–3	7–3	D	= 88 HRS
Lee RN	D	7–3	7–3	7–3	D	7–3	RN 7–7,12	RN 7–7,12	7–3	D	D	7–3	7–3	D	= 80 HRS
Jenny RN	RN 7–7,12	7–7,12	D	D	D	D	D	RN 7–7,12	7–7,12	RN 7–7,12	RN 7–7,12	7–7	D	RN 7–7,12	= 84 HRS
David RN	D	D	RN 7–7,12	RN 7–7,12	RN 7–7,12	RN 7–7,12	D	D	RN 7–7,12	7–3	7–3	D	D	D	= 76 HRS
Cathy	7–3	D	D	7–3	D	D	7–3	7–3	7–3	D	D	7–3	7–3	7–3	= 56 HRS

616 HRS

Legend
- **D** Day off
- **V** Paid vacation day

Handwritten annotations:
- Can work 9 weekends — Carol, Beth
- 9. other weekend off — Steve, Karen, Lee, Jenny, David
- need 5 RN's for day shift
- need a RN to work 7–7
- can work 9 weekend — Stefan, Celia
- 40 M / D

Analysis

1. There appears to be some difficulty with creation of an appropriate schedule. As the manager of this unit, how would you address this issue?

2. If one of the nurses requests to take vacation during this two weeks, what are your options to provide coverage? Discuss the pros and cons of your choices.

Further Reading

Wilson, K. K. (1995). Automated scheduling. *Pennsylvania Nurse*, 50(6), 6–7.

Handwritten at bottom:
avg nursing care hours × days in staffing period × average pt. census.
───
hours of work per FTE in 2 wks

MODULE 5: IMPROVING QUALITY

Exercise 5-1	Case Scenario: Application of the CQI Process
Exercise 5-2	Defining Quality
Exercise 5-3	Comparing CQI and Quality Assurance

Exercise 5-1 Case Scenario: Application of the CQI Process

INTRODUCTION

The continuous quality improvement (CQI) process of the total quality management (TQM) movement has become a central problem of identification and problem-solving method in health care today. TQM as a philosophy requires a commitment by the organization to (1) possess a customer/client focus, (2) dedicate total organization involvement, (3) use quality tools and statistics for measurement, and (4) commit to the identification of key processes and continuous quality improvement. A tool known as a fishbone diagram is used to help understand problems. A fishbone diagram identifies the relationships and possible root causes of an identified problem. As an example, if the nurse manager identified an increase in back injuries of employees during the last 6 months, the diagram might look like the following.

Increased back injuries —— caused by:
- poor lifting technique
- inadequate access to Hoyer lift
- lack of back safety equipment
- increased number of heavy patients

The "bones" of the fishbone diagram are the answers to "why is this occurring?" More bones are added when one continues to ask why each of the original responses occurs. Eventually, one arrives at root causes that can be addressed by the organization.

EXERCISE

Type of Activity: Small or Large Group

Instructions

1. Read the scenario.
2. Create a cause and effect (fishbone) diagram using your imagination and existing knowledge of causes for delays in diagnostic services.
3. As you begin to identify possible causes of delays, you will begin to create the "fishbone" diagram by asking *Why?* repeatedly to each level of the problem. Continue to ask *Why?* until the class agrees that they are at the root of the problem(s).

Scenario

You are the manager for an ambulatory care center that provides outpatient diagnostic and minor emergency services. The center is staffed with nurses, allied health personnel, secretarial staff, and physicians. A diverse range of services is provided, requiring interdependence among all members of the care team. The center is open from 9 AM to 11 PM daily. As the manager of the center, you are aware that patients have complained about the amount of time they wait for results of diagnostic tests. Based upon your knowledge about the services provided, you realize that this complaint does not have a simple answer. To provide continuous improvement in your care delivery, you decide to create a quality team to evaluate this problem.

Discussion

1. Who would you include on the quality team?
2. As you diagrammed this issue, how many different sources of the problem did you identify?
3. Where would you go from here?
4. What are the implications of creating a fishbone diagram of a problem?

Further Reading

Schroeder, P. (1994). *Improving Quality Performance.* St. Louis, MO: C. V. Mosby.

Exercise 5-2 Defining Quality

INTRODUCTION

Total quality management has clarified that every organization must strive for quality. One of the pioneers of the quality movement is Edward Deming. Deming's 14 key points to remember when implementing TQM are listed below. But what does quality really mean? Quality is not a new term for health care; since Nightingale, nursing has emphasized quality in health care.

Deming's 14 Points

1. Create constancy of purpose for improvement of product and service.
2. Adopt the new philosophy.
3. Cease dependence on mass inspection.
4. End the practice of awarding business on the basis of price tag.
5. Improve constantly and forever the system of production and service.
6. Institute training.
7. Institute leadership.
8. Drive out fear.
9. Break down barriers between staff areas.
10. Eliminate slogans, exhortations, and targets for the workforce.
11. Eliminate numerical quotas.
12. Remove barriers to pride of workmanship.
13. Institute a vigorous program of education and retraining.
14. Take action to accomplish the transformation.

Source: Deming, W. E. (1986). *Out of the Crisis.* Cambridge, MA: Massachusetts Institute of Technology.

EXERCISE

TYPE OF ACTIVITY: INDIVIDUAL

Instructions

Answer the following.

1. How would you describe quality? What terms describe quality health care?

2. Who are the key customers of health care?
3. How would the customer(s) describe quality health care?
4. As a nurse manager, how would you determine whether quality care was being given by an unlicensed assistive personnel?

Analysis

1. Are the answers to question number one the same or different? Why?
2. What are the implications for nurse managers?

Further Reading

Berwick, D. M., Godfrey, A. B., & Blanton, J. R. (1990). *Curing Health Care: New Strategies for Quality Improvement.* San Francisco: Jossey Bass.

Deming, W. E. (1986). *Out of the Crisis.* Cambridge, MA: Massachusetts Institute of Technology.

Exercise 5-3 Comparing CQI and Quality Assurance

INTRODUCTION

Health care organizations continue to approach quality issues from two perspectives: continuous quality improvement (CQI) and quality assurance (QA). Both are useful to address problems occurring in health care. The two methods, however, are quite different. Table 1 identifies the differences.

TABLE 1 COMPARING QUALITY ASSURANCE AND CONTINUOUS QUALITY IMPROVEMENT

Quality Assurance	Continuous Quality Improvement
▪ Inspects for quality; reactive	▪ Prevents quality problems; proactive; builds in quality up front
▪ Searches for problem people	▪ Searches for problem processes
▪ Satisfied to meet standards	▪ Never satisfied, always seeking improvement
▪ Clinical people involved	▪ Everybody involved
▪ Lots of required paperwork	▪ Lots of problem solving and experimentation

Adapted from M. W. Koch & T. M. Fairly. (1993). *Integrated Quality Management: The Key to Improving Nursing Care Quality.* St. Louis: CV Mosby.

EXERCISE

TYPE OF ACTIVITY: SMALL GROUP

Instructions

1. Identify a quality problem in health care by group consensus.
2. Divide into small groups, equally divided between QA and CQI approaches.
3. Solve the problem using the two different approaches.

Discussion

1. How would each approach solve the problem?
2. What are the pros and cons of the solutions? Are there situations where one method is more favorable than the other?

Further Reading

O'Leary, D. (1991). CQI—A step beyond QA. *Quality Review Bulletin,* 17(1), 4.

Koch, M. W. & Fairly, T. M. (1993). *Integrated Quality Management: The Key to Improving Nursing Care Quality.* St. Louis: C. V. Mosby.

MODULE 6 — USING COMPUTERS

Exercise 6-1	Case Scenario: Letting Technology Work for You
Exercise 6-2	Understanding Computer Terminology
Exercise 6-3	Accessing the World Wide Web

■ Exercise 6-1 Case Scenario: Letting Technology Work for You

INTRODUCTION

Technology is a major reason for the rapidly changing health care environment. While nurse managers are comfortable with the use of technology in the delivery of nursing care, many are not as comfortable with the use of technology to better manage other resources. Many technologies center around the use of computers, such as electronic mail, word processing, database management, patient classification (acuity) systems, and information retrieval systems such as the World Wide Web. As health care systems incorporate these technologies, nurse managers are challenged to use them to increase their productivity.

EXERCISE

TYPE OF ACTIVITY: INDIVIDUAL

Instructions

1. Read the scenario.
2. Answer the analysis questions.

Scenario

You are the director of rehabilitation services at a large metropolitan hospital. Under your management is a 35-bed rehabilitation unit, physical and occupational therapy services, and the occupational health program targeted at local businesses. Your services are rapidly expanding due to increased marketing efforts in the community. Productivity continues to be a primary issue due to external financial pressures from payors. Your hospital has recently installed computers in the units and has connected them in a local area network (LAN) configuration. Most common software applications are available or can be obtained if justified.

Analysis

1. As the manager, how will you utilize these technological opportunities to improve your management?
2. What opportunities are available to improve staff development?
3. What challenges will occur with computerization? How will you provide ongoing education?
4. What type of back-up system will be necessary?
5. How will the staff access the computers to accomplish charting in a timely fashion?

Further Reading

Brunner, B. K. (1993). Health care-oriented telecommunications: The wave of the future. *Topics in Health Information Management,* 14(1), 54–61.

Gross, M. S., Hoehn, B. J., & Rooks, C. S. (1993). Clinical information systems: Why now. *Topics in Health Information Management,* 14(1), 1–11.

Exercise 6–2 Understanding Computer Terminology

INTRODUCTION

To become computer literate, one must become familiar with terms commonly used to describe components of computerized information systems. As a computer user, increasing one's exposure to terms increases one's opportunity to learn about new technologies.

EXERCISE

TYPE OF ACTIVITY: INDIVIDUAL

Instructions

1. Identify and define the different terms listed below.
2. Explain how these terms can be used in managing nursing care.

   DOS

   LAN

   CPU

   RAM

   ROM

   Netscape

   Medline

   MIS

   WAN

   WWW

   HIS

   CHIN

   NMDS

   UNIX

   GRASP

   Mosaic

Analysis

1. What are some issues that a nurse manager must be concerned about with the explosion of information systems?

Further Reading

Mills, M. E. & Staggers, N. (1994). Nurse-computer performance: Considerations for the nurse administrator. *Journal of Nursing Administration,* 24(11), 30–35.

■ Exercise 6–3   Accessing the World Wide Web

INTRODUCTION

In increasing numbers, nurse managers use the World Wide Web to access information. The World Wide Web is one portion of the Internet that provides a multimedia opportunity to access text, audio, and video information. The Internet is a global network of computers that began as a security system and grew exponentially in the 1980s and early 1990s. Many individuals access the Internet via computer services such as America Online, Compuserve, or Prodigy. Others access the Internet through business local area networks (LANs). To access the Web, however, most individuals are choosing to use a browser interface program such as Mosaic or Netscape (actually called Netscape Navigator). In order to participate in this exercise, either you or your school must have access to the World Wide Web.

EXERCISE

TYPE OF ACTIVITY: INDIVIDUAL

Instructions

1. In the Program Manager of the Local Area Network of your school's computer system, identify the browser interface program icon and double click on it. Most commonly, the interface program will be Netscape. The netsite address for the Netscape Welcome Page is http://home.netscape.com/.

2. Scroll down and read the information presented on the Welcome to Netscape home page.

3. When the Netscape home page appears, look in the toolbar located at the top of your page. You should see a "button" titled Net Search; click on it. If there is no button, scroll down to the bottom of the Netscape home page and and click on the words Net Search (underlined) under the section "Exploring the Net."

    Note: Underlined words are the way one identifies a hypertext link. A hypertext link means that those underlined words are connected to another document. When you click on the hypertext link, you are automatically sent to the next document.

4. The Net Search page (http://home.netscape.com/escapes/internet_search.html) links you to a number of well-known *search engines*. Search engines are online services that literally search the Web for any type of information you could imagine. Yahoo! and Infoseek are two of the more popular search engines.

On the Net Search page, you will see a listing of all the search engines that you have access to from this site. Click on one of the search engine links. The search engine should have a box (usually entitled "Search for:") for you to input a key word. Put your cursor in the textbox and type in "Nursing."

5. The search engine will search the World Wide Web for any sites containing the word "Nursing." You should have an extensive list of sites to choose from. Each of these listings is a hypertext link to another page. Look through these links and find one that interests you. Click on it.

6. To return to a previous document, you may use the Back icon, or open the Go menu and choose the Back command.

7. Document your search each step of the way and bring the results of your search back to class.

8. Should you wish to save a file, you can download a file onto a disk. Most commonly, saving a file requires these steps: open the File menu, click on Save As. If you are in Windows, the option for PCs is Save File As Type and for Macintoshes the option is Format. Your next choice is to select Text or Source; in most cases you will select Source. Save the file with a name and a .HTM extension (example, NURSES.HTM).

9. You can also save the address to links that particularly interest you by choosing "Bookmarks." Just choose "add bookmark" from the bookmark menu while in the particular site that you would like to save, and the address will be kept in your personal list of bookmarks (this is especially beneficial if you are working from your own computer at home so that you can link back to these sites quickly and easily without having to write down the address).

Analysis

1. What was your experience on the World Wide Web?
2. What is the significance of using such an information source?
3. What are the concerns of using the World Wide Web as an *accurate* source of information?

Further Reading

Hoffman, P. E. (1995). *Netscape and the World Wide Web for Dummies*. Foster City, CA: IDG Books Worldwide.

SECTION III
THINKING CRITICALLY

MODULE 7   **DECISION MAKING**

Exercise 7–1   Decision Making Under Uncertainty or Risk
Exercise 7–2   Group Decision Making: The NASA Experience

MODULE 8   **PROBLEM SOLVING**

Exercise 8–1   Learning to Solve Problems

MODULE 9   **CRITICAL THINKING**

Exercise 9–1   Developing Critical Thinking Skills
Exercise 9–2   Using Statistical Tools

| MODULE 7 | DECISION MAKING |

Exercise 7–1   Decision Making Under Uncertainty or Risk

Exercise 7–2   Group Decision Making: The NASA Experience

■ Exercise 7–1 Decision Making Under Uncertainty or Risk

INTRODUCTION

Decision making is a step in the problem-solving process. It may be done singly or in groups. Decision making is defined as the process of establishing criteria by which alternative courses of action are developed and selected.

Types of decisions include routine, adaptive, and innovative. Adaptive and innovative decisions are made when problems and solutions are unclear. The conditions and outcomes of decision making can vary from being *certain* to being *uncertain* and may contain *risk*. When decisions are made under conditions of uncertainty or risk, one must be more concerned about the process of decision making.

Nurse managers can approach decision making from an optimizing or satisfising approach. An *optimizing strategy* involves identifying all possible outcomes and then choosing the action that yields the highest probability of achieving the most desirable outcome. When one uses a *satisfising approach,* one chooses an alternative that is not ideal but is either good enough under existing circumstances to meet minimum standards of acceptance or is the first acceptable alternative.

When one makes a decision involving uncertainty or risk, an optimizing approach is best. Examples of decision-making techniques include the maximax approach, the maximin approach, or the risk-averting approach. All three of these techniques involve predicting outcomes of the decision. The *maximax* approach is to select an alternative whose best possible outcome is the best possible outcome for all alternatives. The *maximin* approach is to compare the worst possible outcome for each alternative and choose the one that seems the least objectionable. Finally, the *risk-averting* approach chooses the alternative with the least variation among all possible outcomes.

EXERCISE

TYPE OF ACTIVITY: SMALL GROUP

Instructions

1. Divide into groups as assigned.
2. Read the assigned scenario.
3. Using the assigned approach, decide how to solve the problem.

Scenarios

1. As the manager of a 12-bed ICU, you recognize that you are seven nursing care hours short for the 3–11 shift. You need to decide what to do to cover the evening shift.
2. You are informed by management that the funds for hiring the open position have been frozen. You must decide whether you will grant the vacation request of a senior staff member when you are already short-staffed.
3. A staff member reported suspicious behavior of an employee that led her to believe that the employee had falsified a record. You must decide how to handle this report.

Discussion

1. In each of the scenarios, what are the risk(s) and/or uncertainties?
2. Using the technique assigned, identify the possible outcomes and the chosen decision. Be prepared to support your choice.
3. What are the strengths and weaknesses of your technique?

Further Reading

Beach, L. R. & Potter, R.E. (1992). The pre-choice screening of options. *Acta Psychologica*, 81, 115–126.

Exercise 7–2 Group Decision Making: The NASA Experience

INTRODUCTION

The widespread use of participative management, quality improvement teams, and shared governance in health care organizations requires every nurse manager to determine when group, rather than individual decision making is desirable and how to use groups effectively. Compared to individual decision making, groups can provide more input, often produce better decisions, and generate more commitment.

EXERCISE

TYPE OF ACTIVITY: SMALL GROUP

Instructions

1. You have been stranded on the lighted side of the moon with other astronauts. In anticipation of rescue, you must prioritize from one to fifteen what items to take with you as you abort the lunar module. Individually complete the NASA Exercise Worksheet (Figure 1).

FIGURE 1 NASA Exercise Worksheet

Individual	Group	
____	____	Box of matches
____	____	Food concentrate
____	____	50 feet of nylon rope
____	____	Parachute silk
____	____	Portable heating unit
____	____	Two .45 caliber pistols
____	____	One case dehydrated milk
____	____	Two 100-pound tanks of oxygen
____	____	Stellar map (of moon's constellation)
____	____	Life raft
____	____	Magnetic compass
____	____	5 gallons of water
____	____	Signal flares
____	____	First-aid kit containing injection needles
____	____	Solar-ed FM receiver-transmitter

2. Upon completion of the individual worksheets, get into groups as assigned.

3. Read the following information before beginning the group process.

   This is an exercise in group decision making. You are to employ a group consensus method in reaching its decision. This means that prioritization of each of the 15 survival items must be agreed upon by each group member before it becomes a part of the group decision. Consensus is difficult to reach. Therefore, not every ranking will meet with everyone's complete approval. Try, as a group, to make each ranking one with which all group members can at least partially agree. Here are some guides to use in reaching consensus:

   a. Avoid arguing for your own individual judgments. Approach the task on the basis of logic.

   b. Avoid changing your mind in order to reach agreement and avoid conflict. Only support solutions with which you are able to at least somewhat agree.

   c. Avoid "conflict-reducing" techniques such as majority vote, averaging, or trading in reaching your decision.

   d. View differences of opinion as helpful rather than as a hindrance in decision making.

4. The group must designate one individual to be the group recorder. The group recorder will assume responsibility for documenting the group's score.

5. Obtain scoring criteria from instructor, and score individual and group priorities as instructed.

Discussion

1. What did you learn about group decision making?
2. What are the implications of group decision-making as compared to individual decision making?

Further Reading

Schoonover-Shoffner, K. (1989). Improving work group decision-making effectiveness. *Journal of Nursing Administration,* 19(7), 10–16.

MODULE 8 — PROBLEM SOLVING

Exercise 8–1 Learning to Solve Problems

Exercise 8–1 Learning to Solve Problems

INTRODUCTION

While problems can be solved using trial-and-error, experimentation, or intuition, a systematic approach to problem solving is useful and efficient. The seven-step method of problem solving to be used in this exercise follows.

Seven Steps to Problem Solving

1. Define the problem.
2. Gather information.
3. Analyze the information.
4. Develop solutions.
5. Make a decision.
6. Implement the decision.
7. Evaluate the solution.

EXERCISE

TYPE OF ACTIVITY: SMALL GROUP

Instructions

1. In your assigned group, use a cause-and-effect (fishbone) diagram (see exercise 5–1) to define an agreed-upon recurrent problem in nursing.

Discussion

1. How did you define the problem(s)?
2. What decision approach did the group choose?
3. What is your solution to this problem?
4. What difficulties did you encounter in using this seven-step process?
5. What are your perceived advantages and disadvantages of group problem solving?

Further Reading

Guyton-Simmons, J. & Ehrmin, J. T. (1994). Problem solving in pain management by expert intensive care nurses. *Critical Care Nurse,* 14(5), 37–44.

MODULE 9 — CRITICAL THINKING

Exercise 9–1	Developing Critical Thinking Skills
Exercise 9–2	Using Statistical Tools

■ Exercise 9–1 Developing Critical Thinking Skills

INTRODUCTION

Critical thinking is an important skill for a nurse manager. *Critical thinking* is a process by which one purposefully engages in mental activity to achieve a specified decision or judgment. Critical thinking requires discipline in the approach to thinking about a problem; however, it does not require one to be rigid. Wilkinson (1995) identified eight characteristics of critical thinking that help to identify when one is thinking critically.

1. Involves conceptualization; of forming a concept.
2. Is rational and reasonable.
3. Is reflective.
4. Is an attitude of inquiry.
5. Is autonomous thinking.
6. Includes creative thinking.
7. Is fair thinking; removes bias.
8. Focuses on deciding what to believe or do.

From Wilkinson, J. M. (1996). *Nursing Process: A Critical Thinking Approach.* Redwood City, CA: Addison-Wesley Nursing.

EXERCISE

TYPE OF ACTIVITY: INDIVIDUAL

Instructions

1. Identify a recurrent problem in nursing not previously discussed by the class.
2. Individually, think about the problem using the eight characteristics listed in the introduction and come up with a solution.

Analysis

1. How do you define the problem as a concept? (conceptualization)
2. What are the facts associated with the problem? (rationality, reflection, inquiry)
3. How did you approach the problem and your solution? (autonomous thinking, creativity)
4. How were you able to put aside your personal bias? (fair thinking)
5. What did you decide? (focused on decision)

Further Reading

Case, B. (1994). Walking around the elephant: A critical thinking strategy for decision making. *Journal of Continuing Education in Nursing,* 25(3), 101–109.

Wilkinson, J. M. (1995). *Nursing Process in Action: A Critical Thinking Approach.* Redwood City, CA: Addison-Wesley.

Exercise 9–2 Using Statistical Tools

INTRODUCTION

The continuous quality improvement process uses statistical methods to identify sources of and solutions to problems. The techniques are commonly referred to as statistical process control (SPC) techniques. Each technique is a different approach to identifying different components of problems—problem identification, sources of problems, or sources of variation. For example, a control chart is used to identify the "normal" amount of expected variance with an activity by timing the activity, developing an average time for the activity, and then comparing subsequent times to the average. Once the norm is established, a nurse manager will recognize when a process is occurring outside of the "normal" range around the average. Occurrences outside the normal range are called *outliners*.

EXERCISE

TYPE OF ACTIVITY: SMALL GROUP

Instructions

1. Pair up as instructed.
2. Time each other (in seconds) while creating a paper airplane.
3. Be prepared to report each individual's time.

Discussion

1. What do you see when you plot out all of the values?
2. How would you determine when an outcome was outside the normal range of variance?
3. How can this information be helpful to a nurse manager?

Further Reading

Schroeder, P. (1994). *Improving Quality and Performance: Concepts, Programs, and Techniques*. St. Louis: Mosby.

SECTION IV: DEVELOPING RELATIONSHIPS

MODULE 10 COMMUNICATION

Exercise 10–1 Evaluating Personal Communication Techniques
Exercise 10–2 The Importance of Effective Organizational Communication
Exercise 10–3 Communicating Effectively Through Writing
Exercise 10–4 Recognizing Communication Patterns
Exercise 10–5 Developing Assertiveness

MODULE 11 GROUPS AND TEAMS

Exercise 11–1 Differentiating Roles People Assume in Groups
Exercise 11–2 Identifying the Stages of Group Development

MODULE 12 MANAGING CONFLICT

Exercise 12–1 Determining When to Intervene in a Conflict
Exercise 12–2 Using Negotiation
Exercise 12–3 Recognizing Role Conflicts

MODULE 10: COMMUNICATION

Exercise 10–1	Evaluating Personal Communication Techniques
Exercise 10–2	The Importance of Effective Organizational Communication
Exercise 10–3	Communicating Effectively Through Writing
Exercise 10–4	Recognizing Communication Patterns
Exercise 10–5	Developing Assertiveness

■ Exercise 10–1 Evaluating Personal Communication Techniques

INTRODUCTION

One of the most important skills in nursing and in management is communication. *Communication* is an ongoing process in which a message is shared between a sender and receiver during an interaction. When messages are shared verbally, a number of nonverbal cues are also given. The recipient of the message must not only attend to the words shared but to the nonverbal cues. Often it is the nonverbal cues that account for the message's impact. Nonverbal cues may be classified as paralanguage (voice tone, pitch, intensity, fluency), visual (eye contact or movement; head or facial agreement or disagreement; shoulder, arm, hand, or finger gestures; body movement, position, or posture; dress and appearance), and touch. *Intrasender conflict* occurs when the verbal and nonverbal messages are incongruent. *Intersender conflict* can occur if conflicting messages are received from different individuals.

The recipient also has a role in enhancing communication. *Active listening* is a technique that not only involves concentrating on what is being said but also involves verbal and nonverbal components. Nonverbal cues include behaviors such as comfortable, but direct, eye contact; an open, relaxed body posture; maintenance of personal space (a distance of about three feet); removal of environmental distractions and physical barriers; and empathy, which is demonstrated by reflecting interest in and concern for the other person. Door openers, open questions, paraphrase, clarification and confirmation are verbal techniques used in active listening. *Door openers* are phrases that invite others to speak such as, "tell me all about. . . ." *Open questions* are inquiries that demand more than single-word answers. An example would be, "What is your opinion about. . .?" Both feeling and content can be paraphrased by summarizing without interpreting what the speaker says or how he or she feels, for example, "You feel frustrated because you can't rely on the nurse extender, Jane, to complete her work assignments." *Confirmation* is used after paraphrasing to determine the accuracy of the paraphrase. *Clarification* is a technique that facilitates understanding vague or uncertain statements. For instance, "I don't understand what you mean by being dumped on."

EXERCISE

Type of Activity: Small Group

Instructions

1. Divide into groups as instructed.
2. In your group, role play the assigned scenario.
3. Use the communication evaluation tool in Table 1 to identify the techniques used and critique the conversation.

Scenarios

1. A staff nurse discusses why she did not get the time off she requested with her nurse manager. The staff nurse has already made plane reservations for which she cannot get her money back.
2. A staff nurse discusses with the charge nurse why she is being reassigned to another unit in which she does not feel comfortable working.
3. A staff nurse discusses with unlicensed personnel why work has not been completed as assigned.
4. A staff nurse discusses with a physician her assessment that the patient is not ready to go home from the hospital.
5. A staff nurse discusses with laboratory personnel an improper technique used to obtain blood via heel stick from a newborn.
6. A staff nurse discusses with a parent the rationale for not bringing children with chickenpox onto the pediatric unit.

TABLE 1	COMMUNICATION EVALUATION TOOL					
Communication Technique	Staff Nurse	Other	Communication Technique	Staff Nurse	Other
Eye contact			Questioning		
Eye movements			Paraphrasing		
Facial expression			Summarizing		
Gestures			Focusing		
Arm movements			Personalizing		
Hand movement			Providing information or giving advice		
Finger movement			Initiating comments		
Fidgeting			Stating feelings		
Inflections in voice			Giving feedback		
Changes in voice quality			Pointing out implicit assumptions		
Touch			Accepting feedback		
Body posture			Repeated assertion		
Silence			Monopolizing the conversation		
Use of timing			Interrupting		
"I" messages			Speaking for others		
Restating					
Reflecting					
Clarifying					

7. A staff nurse discusses with the resident physician why performing a lumbar puncture without permission is a violation of procedure and patient rights.

8. A staff nurse discusses with a patient the need to leave the side rails up after giving preoperative medication.

9. A staff nurse discusses with the director of personnel his disappointment about a change in benefits.

10. A staff nurse discusses with a patient's wife why she cannot spend the night with the patient in his room.

Discussion

1. What techniques did you use the most?
2. Were you unaware of any of the techniques you used?
3. How would you change the way you communicated in this exercise?
4. Which techniques do you want to continue to use?

Further Reading

Axtell, R. E. (1991). *Gestures: The Do's and Taboos of Body Language Around the World*. New York: John Wiley & Sons.

Brown, S. J. (1994). Communication strategies used by an expert nurse. *Clinical Nursing Research*, 3, 43–56.

Ingram, C. A. & Siantz, M. L. (1991). How can we become more aware of culturally specific body language and use this awareness therapeutically? *Journal of Psychosocial Nursing Mental Health Services*, 29(11), 38–41.

Exercise 10-2 The Importance of Effective Organizational Communication

INTRODUCTION

In organizations, communication may be downward (management to staff), upward (staff to management), lateral (between individuals at the same hierarchical level), or diagonal (individuals/departments at different levels in the hierarchy). The effectiveness of the communication, whether formal or informal, is dependent upon six components: accessibility of information, communication channels, organizational structure, clarity of message, flow control and information load, and communicator effectiveness. Communicator effectiveness is dependent upon the manager's communication skills. Accessibility of information can influence job performance. Is the predominant communication channel formal or informal? How long does it take for information to travel? Who are the key people? Is the structure centralized, which facilitates communication, or decentralized? Sending clear messages, as well as sharing the appropriate amount of information, promote trust. Restricting information threatens productivity and encourages rumors, whereas providing too much information produces overload, which limits the processing of information and increases errors.

EXERCISE

Type of Activity: Small or Large Group

Instructions

1. Get into groups as assigned and obtain written instructions and balloons from the instructor.
2. When instructed, blow up as many balloons as possible in a 5-minute time period.

Discussion

1. Did you have all the information you needed? If not, what information would you have liked? Did you have too much information?
2. How was productivity, satisfaction, and group effectiveness affected? Which group was the most productive?

Further Reading

Conrad, C. (1990). *Strategic Organizational Communication*. Fort Worth: Holt, Rinehart & Winston.

Farley, M. J. (1989). Assessing communication in organizations. *Journal of Nursing Administration*, 19(12), 27–31.

Exercise 10-3 Communicating Effectively Through Writing

INTRODUCTION

Writing is another skill needed by professionals and managers. Communicating effectively through writing takes practice. Although lack of immediate feedback is a problem inherent in written communication, written formats provide documentation and are ideal for routine and simple communication. In writing memos, letters, minutes, or other documents, certain considerations are important. First of all, written documents represent the writer and the organization to the receiver. So consider the form, or how you will write. Neatness, organization, clarity, and tone are all important. Decide what it is that you want to say before you start writing. Who are you writing to and what or who are you writing about? Be sure to include why you are writing. Arrange the information logically. Keep statements positive. Use as few words as possible in simple, direct sentences; only introduce one idea per sentence. Support your position with facts. Connect your thoughts with selected transitional words.

EXERCISE

TYPE OF ACTIVITY: INDIVIDUAL

Instructions

Pretend you are the nursing manager of a 12-bed intensive care unit. Select and write one of the following:

a. A memo to the department head of dietary. Describe a continuous problem of cold trays due to dietary assistants bringing up tray carts and leaving them in the hall for nursing to deliver, but not informing nursing of their presence.

b. A report to administration explaining variance in the personnel budget due to high acuity patients.

c. A brochure describing a new technology available in your unit.

d. A news release about a new program offered by your department.

Analysis

1. Why did you pick the project you did?
2. What difficulties did you have in completing the project?

Further Reading

Schneller, G. & Godwin, C. (1983). *Writing Skills for Nurses: A Practical Text/Workbook*. Reston, VA: Reston.

Wallace, H. (1989). How to improve your written communication. *Medical Laboratory Observer*, 21(3), 65–68.

Exercise 10-4 Recognizing Communication Patterns

INTRODUCTION

A variety of communication networks, depicted in Figure 1, develop in groups and organizations. The effectiveness of these patterns, the emergence of leaders, power bases, and member satisfaction also vary. In the *chain* network, leadership is centralized, communication patterns are fixed, and morale is low. However, communication is fast and leadership is stable. The *y, star,* and *wheel* patterns also have centralized leadership which yield fast, efficient communication, but again members are less satisfied. Both the circular and all-channel patterns represent decentralized leadership. The circular structure, though slow, is flexible and enhances staff morale. The all-channel pattern is best for solving complicated problems.

FIGURE 1 Types of Communication Networks

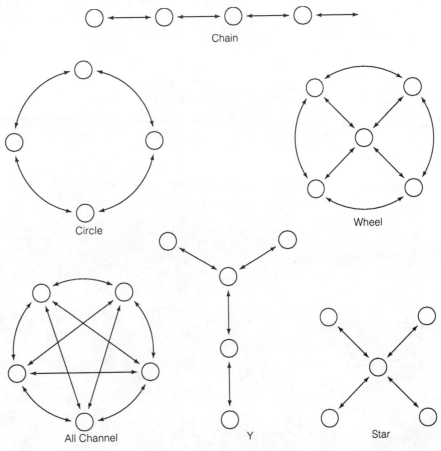

From Bass, B. M. (1990). *Bass & Stogdill's Handbook of Leadership: Theory, Research, and Managerial Applications.* (3rd edition). New York: Free Press.

EXERCISE

TYPE OF ACTIVITY: SMALL OR LARGE GROUP

Instructions

1. Divide into groups as instructed and obtain directions from the instructor.
2. Select an individual in each group to serve as an observer-recorder. The observer-recorder should track the path of communication within the group.

Discussion

1. How fast was communication in the group?
2. How satisfied was the group with the communication process?
3. What type of communication network was operating?

Further Reading

Bass, B. M. (1990). *Bass & Stogdill's Handbook of Leadership: Theory, Research and Managerial Applications*. (3rd ed.). New York: Free Press.

■ Exercise 10–5 Developing Assertiveness

INTRODUCTION

Communication may be nonassertive, aggressive, or assertive. These are compared in Table 1. Passive communication is often used to avoid unpleasant situations such as conflict and confrontation. It is an inefficient form of communication often resulting in low self-esteem, self-pity, feelings of powerlessness, anxiety, frustration, and anger. Characteristics of passive communication are apologetic speech, avoidance,

TABLE 1 A COMPARISON OF NONASSERTIVE, ASSERTIVE, AND AGGRESSIVE COMMUNICATION

	NONASSERTIVE	ASSERTIVE	AGGRESSIVE
VERBAL	■ Apologetic words. Veiled meanings. Hedging; failure to come to the point. Rambling; disconnected. At a loss for words. Failure to say what you really mean. Qualifying statements with "I mean," "you know."	■ Statement of wants. Honest statement of feelings. Objective words. Direct statements that say what you mean. "I" messages.	■ "Loaded" words. Accusations. Descriptive, subjective terms. Imperious, superior words. "You" messages that blame or label.
NONVERBAL			
■ General demeanor	■ Actions instead of words, hoping someone will guess what you want. Looking as if you don't mean what you say.	■ Attentive listening behavior. Generally assured manner, communicating caring and strength.	■ Exaggerated show of strength. Flippant, sarcastic style. Air of superiority.
■ Voice	■ Weak, hesitant, soft, sometimes wavering.	■ Firm, warm, well-modulated, relaxed.	■ Tense, shrill, loud, shaky; cold, "deadly quiet," demanding; superior, authoritarian.
■ Eyes	■ Averted, downcast, teary, pleading.	■ Open, frank, direct. Eye contact, but not staring.	■ Expressionless, narrowed, cold staring; not really "seeing" others.
■ Stance and posture	■ Leaning for support, stooped, excessive head nodding.	■ Well-balanced, straight on, erect, relaxed.	■ Hands on hips, feet apart. Stiff and rigid. Rude, imperious.
■ Hands	■ Fidgety, fluttery, clammy.	■ Relaxed motions.	■ Clenched. Abrupt gestures, finger-pointing, fist pounding.

Source: E. Zuker (1993). *Mastering Assertiveness Skills.* New York: AMACOM, a division of American Management Association, New York University, p. 15.

withdrawal, a weak or hesitant voice, downcast eyes, fidgety hands, and frequent nodding. Individuals who use passive communication want to please and as a result are easily manipulated, an undesirable trait in a manager.

Conversely, aggressive individuals may have either high or low self-esteem and self-worth. They get what they want at the expense of others. Aggressive individuals are often direct, loud, sarcastic, rude, and hostile. Characteristic behaviors include blaming, belittling, humiliating, or embarrassing others and assuming an air of superiority. As a result, the individuals targeted by the aggressor are humiliated, hurt, angry, and vengeful.

A passive-aggressive communication style is the delivery of an aggressive message in a passive manner. Individuals using this style often withdraw in an effort to manipulate the situation or send incongruent verbal and nonverbal messages.

The most effective communication style is assertiveness. Assertiveness is the open, honest expression of feelings in a manner respectful of others. Melodie Chenevert (1988) has identified a number of rights and responsibilities of assertive individuals that are listed in Table 2. Common techniques used by assertive individuals include reflection, using "I" messages, active listening, restating, repeated assertion, negative assertion, fogging, broken record, pointing out implicit assumptions, negative inquiry, and questioning. The end result of assertive communication is high self-esteem, good interpersonal relationships, and respect by others and the achievement of goals.

TABLE 2 RIGHTS AND RESPONSIBILITIES OF ASSERTIVE INDIVIDUALS

RIGHTS	RESPONSIBILITIES
■ To speak up	■ To listen
■ To take	■ To give
■ To have problems	■ To find solutions
■ To be comforted	■ To comfort others
■ To work	■ To do your best
■ To make mistakes	■ To correct your mistakes
■ To laugh	■ To make others happy
■ To have friends	■ To be a friend
■ To criticize	■ To praise
■ To have your efforts rewarded	■ To reward others' efforts
■ To be independent	■ To be dependable
■ To cry	■ To dry tears
■ To be loved	■ To love others

Source: Chenevert, M. (1988). *Pro-Nurse Handbook*. (3rd ed.). St. Louis: C. V. Mosby.

EXERCISE

Type of Activity: Small or Large Group

Instructions

1. Divide into groups as instructed. Select one individual in each group to be an observer-recorder.
2. Obtain scenario from instructor. Decide who will role-play each character.
3. Obtain behavior card from instructor.
4. Role-play the assigned behaviors within the time frame allotted.
5. Upon completion of the exercise, address the discussion questions within your own group and then again as a class.

Discussion

1. What type of behavior was displayed by each character?
2. Did anyone have difficulties playing their role and using the behavior assigned?
3. How did the different combinations of behaviors affect communication? What combination was most effective?
4. How could a change in behavior positively affect collaboration or productivity?

Further Reading

Chenevert, M. (1988). *Pro-Nurse Handbook*. (3rd ed.). St. Louis: C. V. Mosby.

MODULE 11: GROUPS AND TEAMS

Exercise 11–1 Differentiating Roles People Assume in Groups

Exercise 11–2 Identifying the Stages of Group Development

■ Exercise 11–1 Differentiating Roles People Assume in Groups

INTRODUCTION

Groups are social systems that evolve informally through commonalities or formally through organizational goals. Groups may accomplish a number of different types of tasks. When the group's performance is dependent on the sum of the individuals' performance, the group is involved in an *additive task*. In *disjunctive tasks,* the group succeeds if only one of its members succeeds. Conversely, in a *conjunctive task* the group succeeds only if all its members succeed. Finally, *divisible tasks* or tasks that can be broken down into subtasks provide an opportunity for specialization and interdependence.

To facilitate the group's performance its members can assume a combination of functional task or socioemotional (nurturing) roles. Task, nurturing, and dysfunctional roles are summarized below.

Task Roles

Initiator-contributor Redefines problems and offers solutions, clarifies objectives, suggests agenda items, and maintains time limits.

Information seeker Pursues descriptive bases for the group's work.

Information giver Expands information given by sharing experiences and making inferences.

Opinion seeker Explores viewpoints that clarify or reflect the values of other members' suggestions.

Opinion giver   Conveys to group what their pertinent values should be.

Elaborator   Predicts outcomes, provides illustrations, or expands suggestions, clarifying how they could work.

Coordinator   Links ideas or suggestions offered by others.

Orienter   Summarizes the group's discussions and actions.

Evaluator–critic   Appraises the quality and quantity of the group's accomplishments against set standards

Energizer   Motivates the group to qualitatively and quantitatively accomplish its goals.

Procedural technician   Supports group activity by arranging the environment (e.g., scheduling meeting room) and providing necessary tools (e.g., ordering visual equipment).

Recorder   Documents the group's actions and achievements.

Nurturing Roles

Encourager   Compliments members for their opinions and contributions to the group.

Harmonizer   Relieves tension and conflict.

Compromiser   Submits own position to maintain group harmony.

Gatekeeper   Encourages all group members to communicate and participate.

Group observer   Takes note of group process and dynamics and informs group of them.

Follower   Passively attends meetings, listens to discussions, and accepts group's decisions.

Dysfunctional Roles

Aggressor   Attacks and criticizes others in an attempt to meet own needs.

Blocker   Inhibits group progress by being resistant, negative, or disagreeable.

Dominator   Attempts to usurp leadership of the group.

Help-seeker   Solicits sympathy from the group through expressions of insecurity, confusion, and ineptness.

Monopolizer   Dominates the conversation prohibiting others from speaking.

Playboy/Playgirl   Belittles seriousness of group work and is nonchalant; plays around, jokes, makes irrelevant and silly comments.

Recognition seeker   Draws attention to self through boasts and acting-out behaviors.

Self-confessor   Uses group for expression of personal feelings.

Special interest pleader   Presents and supports issues relevant to a particular group.

Zipper-mouth   Does not accept or participate in group process; may sulk.

EXERCISE

Type of Activity: Small or Large Group

Instructions

1. Divide into groups as instructed and select an observer-recorder.
2. Obtain a deck of cards from the instructor.
3. In the time allotted, build a tower. All cards must be used, and no other materials can be used.
4. The observer-recorder should note the following:
    a. Who assumed the leader role?
    b. Did the leader evolve, or was the leader appointed?
    c. Did everyone participate in the project?
    d. What group roles were apparent? Use Table 1 to track participant's roles.
5. Upon completion of the exercise the observer-recorder should report the findings.

Discussion

1. What type of group task (additive, disjunctive, divisible, conjunctive) did this exercise represent?
2. How productive was each team?
3. Was the group cohesive in its efforts?
4. Was anyone surprised by the roles they assumed?

Further Reading

Johnson, D. & Johnson, S. (1994). *Joining Together: Group Theory and Group Skills.* (5th ed.). Boston, MA: Allyn & Bacon.

TABLE 1	GROUP ROLES WORKSHEET					
TASK ROLES						
■ Initiator-contributor	all of us.					
■ Information seeker	all of us.					
■ Information giver						
■ Opinion seeker						
■ Opinion giver						
■ Elaborator						
■ Coordinator						
■ Orienter						
■ Evaluator–critic						
■ Energizer						
■ Procedural technican						
■ Recorder	CC					
NURTURING ROLES						
■ Encourager						
■ Gatekeeper						
■ Compromiser						
■ Harmonizer						
■ Follower						
■ Group observer	CC					
DISFUNCTIONAL ROLES						
■ Aggressor						
■ Blocker						
■ Dominator						
■ Help-seeker						
■ Monopolizer						
■ Playboy/Playgirl						
■ Recognition seeker						
■ Self-confessor						
■ Special interest pleader						
■ Zipper-mouth						

Exercise 11–2 Identifying the Stages of Group Development

INTRODUCTION

Groups go through five stages of development: forming, storming, norming, performing, and adjourning. While *forming,* the group relies upon the leader to define its purpose, tasks, and roles. Members contemplate their own contributions, test the boundaries of interpersonal behaviors, and determine group membership. Members are anxious and self-protective.

During the *storming* stage, conflicts avoided while the group was forming emerge and tension is high. Trust is low. Competition for power and status occurs and informal leadership surfaces. The group's leader helps the group to acknowledge conflicts and to channel energies constructively, so that by the end of this stage the group resolves its conflicts and reorganization occurs.

The group develops cohesiveness and establishes functional patterns of behavior during the *norming* stage. Trust, confidence, and loyalty allow the members to set aside their differences and establish goals and rules of behavior. The leader facilitates this process by keeping the group focused and promoting relationship building.

By the time the group reaches the *performing* stage it has defined its purpose, has identified its objectives, and has developed a plan. The group is now task oriented. All members carry out their share of the work and relate effectively to the group and to other members. Cooperation is evident as trust is high and each member's contribution is appreciated. Communication is open, differences are tolerated, and feedback is constructive. Excitement, recognition, and rewards are common. The leader facilitates the group's work by providing feedback on the quality and quantity of work, praising achievements, and reinforcing interpersonal relationships.

Upon completion of its objectives, a group usually disbands or *adjourns.* The task at this time is to evaluate the process and achievements of the group. Emotions are often mixed. The leader not only prepares the members for its dissolution, but also facilitates closure by expressing appreciation and giving positive feedback.

EXERCISE

Type of Activity: Individual

Instructions

1. Attend a health care organization meeting.
2. Be able to answer the following questions:
    a. How long has the group been meeting?
    b. How long has the present leader been in charge?
    c. Is the group stable, or does its membership change frequently?
    d. What is the purpose for the group?

Discussion

1. What stage of development would you evaluate the group to be in? Why?
2. How does the leader interact with the group?
3. What are the group dynamics?

Further Reading

Lacoursier, R. B. (1980). *The Life Cycle of Groups: Group Developmental Stage Theory.* New York: Human Sciences Press.

Tuckman, B. W. (1977). Stages of small group development revisited. *Group and Organization Studies,* 2(4), 419–427.

Tuckman, B. W. (1965). Developmental sequences in small groups. *Psychological Bulletin,* 72, 384–399.

MODULE 12   MANAGING CONFLICT

Exercise 12–1	Determining When to Intervene in a Conflict
Exercise 12–2	Using Negotiation
Exercise 12–3	Recognizing Role Conflicts

■ Exercise 12–1 Determining When to Intervene in a Conflict

INTRODUCTION

Conflict is a dynamic force; the consequence of real or perceived differences in mutually exclusive goals, values, ideas, attitudes, beliefs, feelings, or actions. Conflict may be beneficial or detrimental. Conflict often increases the awareness of an issue, stimulates creativity, and improves performance, effectiveness, and satisfaction. However, conflict may also lead to scapegoating or other disruptive, irrational, or even violent behavior. Different types of conflict have been described. *Competitive conflict* occurs when there are mutually incompatible goals and the emphasis is on winning. *Disruptive conflict* involves activities to reduce, defeat, or eliminate the opponent. *Structural conflict* is the result of poor communication, competition for resources, opposing interests, or lack of shared interests between different levels within the organizational structure.

EXERCISE

TYPE OF ACTIVITY: INDIVIDUAL OR GROUP

Instructions

Select one of the following scenarios and address the questions under "Analysis."

Scenarios

1. Susan and Margarette are both nurses for a home-health care agency. They both started working at the agency the same day, two years ago. Susan received her ADN from the local community college; Margarette has a BSN from the state university. There seems to have been tension between the two of them from day one. You frequently hear Susan complaining about her caseload and comparing it to Margarette's load. You know there is no preferential treatment since cases are assigned as they come in to the next nurse on an alphabetical list. Today you overhear Susan taunt Margarette, "So, are you planning another trip to Jamaica with all the overtime money you made last week?"

2. You overhear Dr. Delgado, one of the surgeons, yelling and throwing things in a patient's room. Mary, one of your staff nurses, is in the room. She calmly asks Dr. Delgado if he would meet her in the hall. He follows her out. She calmly informs him that she doesn't appreciate his behavior and expects him to treat her and the patient with more respect. Before walking off she states, "Furthermore, I will need to report this unprofessional behavior to my manager and the Chief of Surgery."

3. Mary and Tse are two of the best nurses on the unit. Both have applied for vacation time for June 1 through June 8, but only one can take off at a time. You have told them to work it out between them as to who can take vacation time then and get back to you by April 29. It is May 1 and neither one has gotten back to you.

4. Alice and Gary frequently have charge responsibilities on the unit. Because of some miscommunication, both believe they are to be in charge tonight. Since being in charge means getting an added differential, both want the responsibility and are fighting at the nurse's station about who is really in charge.

Discussion/Analysis

1. What is the conflict?
2. What type of conflict (competitive/disruptive/structural) is occurring?
3. What steps are the involved parties taking to resolve their conflict? What technique does this represent?
4. Is intervention (mediation) necessary?
5. If mediation is necessary, identify the conflict handling mode (competition, collaboration, compromise, avoidance, accommodation) you would use.

Further Reading

Barton, A. (1991). Conflict resolution by nurse managers. *Nursing Management,* 22(5), 83–84, 86.

Cavanaugh, S. J. (1991). The conflict management style of staff nurses and nurse managers. *Journal of Advanced Nursing,* 16, 1254–1260.

Martin, K., Wimberly, D. & O'Keefe, K. (1993). Resolving conflict in a multicultural nursing department. *Nursing Management,* 25(1), 49–51.

Exercise 12–2 Using Negotiation

INTRODUCTION

When conflict is negative or dysfunctional, attempts at resolution are necessary. Thomas (1992) has described five approaches to conflict resolution: competing, collaborating, compromising, avoiding, and accommodating. These modes reflect two underlying dimensions: level of assertiveness (satisfaction of personal concerns) and level of cooperativeness (satisfaction of other's concerns). *Competition* is an effort to win regardless of the cost. Assertiveness is high while cooperativeness is low. Although competition may be necessary in situations involving unpopular or critical decisions or when time does not allow for more cooperative techniques, a win-lose situation always occurs. This technique often leaves the loser angry and frustrated.

In *collaboration* both parties' concerns are addressed. Both assertiveness and cooperativeness are high resulting in a win-win solution. The focus is on solving the problem and not on defeating the opponent. Although collaboration is the best conflict management technique, it is not always feasible.

Compromise is also high on assertiveness and cooperativeness, but because of trade-offs the best outcome is a win-lose solution. In compromise, negotiations are made until some middle ground, satisfactory to all involved, is reached. Compromise is the most common conflict management technique used by managers. Common negotiation tactics are to separate the people from the problem, to focus on issues versus different positions, and to identify possible solutions. It is also helpful to identify whether the other party is an ally (high agreement, high trust), opponent (low agreement, high trust), adversary (low agreement, low trust), bedfellow (high agreement, low trust), or fence sitter (medium agreement, low trust). (See Figure 1.) Negotiations differ with each. While doubts and vulnerabilities can be shared with allies, it is important to limit the information shared with bedfellows since trust is an issue. Conversely, opponents can be trusted; therefore, the issue is a difference in positions. Adversarial relationships are the result of previous failures in negotiation. With this group it is important to state your position and your perception of the adversary's position, your contribution to the conflict, and your plans for resolution without making any demands on the adversary. Working with fence sitters is similar, except that you seek a final decision. Express your frustration with the other party's neutrality and attempt to elicit the fence sitter's support.

FIGURE 1 Agreement-Trust Grid

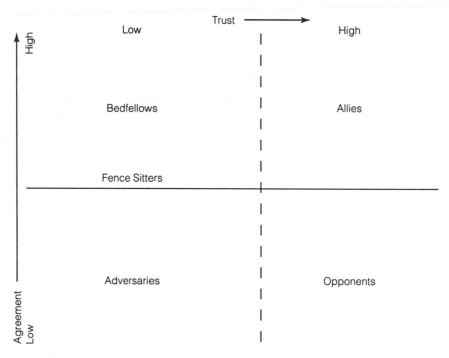

Block, P. (1991). *The Empowered Manager.* San Francisco: Jossey-Bass.

Accommodating is a non-assertive, cooperative tactic used to preserve harmony. It can be an appropriate technique if the accommodating party does not have a vested interest in the issue, but clearly one party wins and the other party loses.

Avoiding is a non-assertive, uncooperative technique that results in a lose-lose situation. Avoidance is often seen in highly cohesive groups. Sometimes avoidance is used when the costs of managing the conflict outweigh the benefits, or when the problem will resolve itself.

EXERCISE

Type of Activity: Individual

Instructions

1. Think of a recent conflict in your life.
2. Using Figure 1, identify whether the individual was an adversary, ally, bedfellow, fence sitter, or opponent.
3. Use the questions under "Analysis" to evaluate the situation.

Analysis

1. What were each individual's issues? Emotions?
2. What positions were taken? What tactics were used?
3. Were you able to negotiate your stand with the other party?
4. What was the outcome of the conflict?
5. Is there anything in retrospect you would do differently?

Further Reading

Block, P. (1991). *The Empowered Manager.* San Francisco: Jossey-Bass.

Levenstein, A. (1984). Negotiation vs. confrontation. *Nursing Management,* 15(1), 52–53.

Thomas, K. W. (1992). Conflict and conflict management: Reflections and update. *Journal of Organizational Behavior,* 13(3), 265–274.

Exercise 12-3 Recognizing Role Conflicts

INTRODUCTION

Roles are socially prescribed expectations for behavior by a person in a particular position. In addition to their professional roles, nurses are often parents, children, and spouses. Multiple roles often lead to multiple demands and *interrole conflict*. When expectations of an individual by two or more parties differ, this leads to *intersender conflict*. *Person-role* conflicts occur when role expectations are incompatible with the person's personality, skills, or abilities (e.g., incompetence, overqualification). Role conflict also occurs when roles are poorly defined (role ambiguity) or too demanding (role overload).

EXERCISE

TYPE OF ACTIVITY: INDIVIDUAL

Instructions

1. List your current roles.
2. Identify what type of role conflict is represented by each of the following scenarios.

Scenario

1. You are asked to work nights.
2. Day shift hours are changed to 6:30 AM to 3:00 PM.
3. You have to work Christmas Day from 11:00 AM to 11:00 PM.
4. You are scheduled to work on your first wedding anniversary.
5. Your car breaks down and you have to take the bus, which is late, making you late for work.
6. Your father comes into the emergency room in a cardiac arrest and you have to care for him as a patient.

Analysis

1. Would any of these scenarios introduce conflict in your life?
2. Which situation would be the most difficult to resolve? Why?
3. How would you go about resolving the conflict?

Further Reading

Benner, P. (1984). *From Novice to Expert*. Menlo Park, CA: Addison-Wesley.

SECTION V: MANAGING TIME AND STRESS

MODULE 13 SETTING PRIORITIES

Exercise 13–1 Setting Priorities: An In-Basket Exercise
Exercise 13–2 Recognizing Time Wasters
Exercise 13–3 Managing Your Professional Time

MODULE 14 DELEGATION

Exercise 14–1 Case Scenario: Use of Delegation
Exercise 14–2 Differentiating Responsibility, Accountability, and Authority

MODULE 15 STRESS ASSESSMENT AND REDUCTION

Exercise 15–1 Reducing Organizational Stress

MODULE 13 — SETTING PRIORITIES

Exercise 13–1	Setting Priorities: An In-Basket Exercise
Exercise 13–2	Recognizing Time Wasters
Exercise 13–3	Managing Your Professional Time

Exercise 13–1 Setting Priorities: An In-Basket Exercise

INTRODUCTION

"I don't have time to do my job" is a typical phrase heard from nurses who often experience their jobs as fragmented blurs of activities. Clearly, one of the hardest tasks for any nurse is to manage time effectively.

To increase personal effectiveness, a nurse must manage his or her time by setting priorities. One way to do this is to make a list of things to be accomplished each day in terms of the following categories.

1. Needs action today.
2. Delegate to someone (specify to whom).
3. Postpone action until specified time.
4. No action needed.

Both the importance and the urgency of the task must be considered when determining what task to work on first. A nurse should avoid waiting until a crisis to do an urgent task and focusing on unimportant tasks at the expense of important ones. Two useful rules to help prioritize and delegate tasks are: (a) if something is not urgent, then it can wait; and (b) if it isn't important, then it can be delegated to someone else.

Listing or gathering 5- to 10-minute discretionary tasks is also a useful time-management technique. These can be used as fillers between important events. It is essential to determine the critical 20 percent of your tasks that produce 80 percent of your results. Spend most of your time on that 20 percent.

Another rule of effective time management is to evaluate the results of your activities rather than the effort expended. Don't continue to procrastinate. Most tasks take less time if they are done earlier. Keeping a time log periodically can help you determine how much time is being spent on which tasks. Setting deadlines and delegating routine tasks also are helpful techniques for time management.

Finally, don't forget to schedule personal time. You need this time in order to plan, meditate, or just relax. This will help you maintain the self-awareness you need in order to be effective.

EXERCISE

Type of Activity: Small Group

Instructions

1. Read the scenario and follow the instructor's directions for completing the in-basket forms.

2. Upon completion of allotted time, divide into groups of four to six members.

3. In small groups, answer the following discussion questions.

Scenario

You are a nurse manager on a 28-bed medical division and have held this position for one year. Your division has one assistant nurse manager, 13 staff nurses, seven LPNs, and three unit secretaries. You have a unit manager responsible for the division's supplies and equipment.

Today is Thursday, June 25. The time is 9:45 AM. Your name is Janet Baccard. You have received a report from the night nurse and have checked division assignments and patient status.

Your assistant is in charge today, so you would like to take advantage of this opportunity to spend a couple of hours catching up on paper work, schedules, evaluations, etc.

Today's schedule includes a policy and procedure committee meeting (1:00 to 2:00), a nurse manager meeting (3:00 to 4:00), and a counseling session at 4:30 with one of your employees.

You are now at your desk preparing to deal with the multitude of letters, memos, etc., you find in your in-basket.

You have 30 minutes available to you to take whatever action you consider necessary on each item in your in-basket.

You have total discretion and responsibility to arrange your day such that priorities (in your judgment) are taken care of. On the form on page 100, the in-basket items listed are in numerical order. Write your decision for handling each item in the space provided, stating whether:

A. you are taking action now (today) and, if so, what action.
B. you are delegating the action to someone else and, if so, to whom.
C. you are postponing action and, if so, for how long.
D. no action is necessary.

The in-basket items appear on pages 101–115.

Discussion

1. How do the group members' lists compare and contrast? What are the main differences?
2. Did members scan all items first or begin working on the first item? Which is a better strategy?
3. How was the decision to delegate certain items made?
4. What are the strengths and weaknesses of each person's list?
5. How were decisions made about time for postponement of tasks?

Further Reading

Covey, S. R. (1990). *The Seven Habits of Highly Effective People: Powerful Lessons in Personal Change.* New York: Simon & Schuster.

Mayer, J. (1990). *If You Haven't Got the Time to Do It Right, When Will You Have the Time to Do It Over?* New York: Simon and Schuster.

IN-BASKET DECISION FORM

NAME _____

In-Basket Item Number	Decision (state alternative, if any)	Briefly explain why the decision was made
MEMO 1		
MEMO 2		
MEMO 3		
MEMO 4		
MEMO 5		
MEMO 6		
MEMO 7		
MEMO 8		
MEMO 9		
MEMO 10		
MEMO 11		
MEMO 12		
MEMO 13		
MEMO 14		

MEMO 1

Janet,

Things are getting increasingly tense with the Madsens. They're complaining about everything we're doing for their mother. They're angry not only about the care but also about the attitude with which the care is delivered.

I feel that the care Mrs. Madsen is receiving is straight out of the handbook, but it is true that the staff's attitude is somewhat less than warm—naturally, no one wants to go in there. Last evening the Madsens did some loud ranting and raving, disturbing patients on both sides of their mother's room. I've not been able to make any impact on them. I suspect that most of this behavior is a projection of the anger they feel over her imminent death. Would you please try to talk to them? It's been very stressful for me as well as the staff. Thanks.

                                                                  Beverly
                                                 (Asst. Nurse Manager)

MEMO 2

Janet,

For the fourth consecutive evening, we've not had an adequate supply of angiocaths and have had the inconvenience of waiting for them. Can't you do something about getting some extras, at least until the census is down?

                                                     Maria
                                              (Staff Nurse, 3–11)

MEMO 3

HOSPITAL NURSING SERVICE

TO: Head Nurses
FROM: Linda Hom, RN, Asst. Director
Nursing Service
DATE: June 15
RE: MD/Nurse Committee Pilot Project

As there has been interest shown by the attending physicians to localize placement of their patients more consistently, the above nursing divisions will be participating in a pilot project beginning July 5.

In order to determine if this type of placement is feasible, it will be necessary for each above division to submit a daily record of patient census and the patient's respective attending physician. (See attached example.)

The "Individual Assignment Sheet" form is to be used, and nursing division specified at top. Please have your Unit Clerk place these in my mailbox no later than 1200 each day. This project will be in effect from July 5 to October 31.

LH/sc
Attachment

MEMO 4 (see attachment on page 106)

June 20

Janet,

Could you quickly fill out this reference form so it won't be late arriving at the school. I would really appreciate *a lot* of good words as I'm so intent on beginning nursing school. Thanks so much.

Linda Thompson
(Nurse aide, 7–3)

MEMO 3 Attachment

NURSING DEPARTMENT
INDIVIDUAL ASSIGNMENT SHEET

Name 6300　　　　　　　　　　　　　　　　　　　　　　　　　Date 5/24

Room No.	Name	NPO	I&O	IV	Special Information	Special Tests and Procedures
34	Bekkar				Dr. Corbett	
B	Koch				Dr. Thomas	
35	Wilson				Dr. Garcia	
36	Nguyen				Dr. Fong	
B						
37	Craft				Dr. Fong	
B	Primas				Dr. Garcia	
38	Webb				Dr. Thomas	
B						
39	Plansky				Dr. Garcia	
40	Lucas				Dr. Corbett	
41	Hollensbe				Dr. Thomas	
42	Atchison				Dr. Corbett	
B	Nelson				Dr. Fong	
43	Kontz				Dr. Garcia	
44						
B	Matsura				Dr. Thomas	
45	Casper				Dr. Johnson	
B	Wilton				Dr. Garcia	
46						
B	Mc Donald				Dr. Fong	
47	Stein				Dr. Corbett	
B	Warthen				Dr. Thomas	
48	Burton				Dr. Corbett	
49	Starling				Dr. Garcia	
B	Haynes				Dr. Thomas	
50						
51	Furlow				Dr. Johnson	

MEMO 4 Attachment

FROM: Janet Baccard, HN

TO: Charlotte Henderson, RN

RE: Linda Thompson's application

The above named person has made application to the Hospital School of Nursing and has given your name as a reference.

Your frank evaluation of the applicant as requested will be greatly appreciated. All information given will be kept confidential.

How long have you known the applicant? _____

What in your opinion are the strengths and weaknesses of the applicant? _____

_____

_____

_____

_____

Would you recommend the applicant for admission? _____

Additional Comments: _____

_____

_____

_____

_____

Signature: _____

MEMO 5

HOSPITAL NURSING SERVICE

TO: All Medicine Nurse Managers
      and Assistant Nurse Managers
FROM: Suzanne Estes
DATE: June 23
RE: Medical Assistant Nurse Manger
      Group Meeting

Please arrange to have your Assistant Nurse Managers attend the first planning meeting for the Medical Assistant Nurse Manager Group on July 12 at 1300–1430 in the 8300 Conference Room.

Thanks

SJ

MEMO 6

HOSPITAL NURSING SERVICE

TO: Janet Baccard
FROM: Linda Hom
      Asst. Director
DATE: June 24
RE: Scheduling

While making rounds Friday evening the RN staff was very verbal about working the 12–hour shift on weekends. They did not seem to have a clear understanding of why they were being "forced" to work such hours or for how long this scheduling would last. I encouraged them to discuss this with you as soon as possible.

MEMO 7 (see Attachments A and B)

INCIDENT REPORT / Patient-Visitor

NOTICE: Consider whether Security should be called immediately, especially in cases of theft or disappearance of patient or employee valuables, hospital property, including narcotics, keys to offices or narcotics cabinets, automobiles and contents thereof, as well as vandalism, fights, or threats of physical harm.

"An incident is any happening which is not consistent with the routine operation of the hospital or the routine care of a particular patient. It may be an accident or a situation which might result in an accident."

INSTRUCTIONS FOR USE

TIMELY SUBMISSION

The last part of form (White Sheet—front and back) must be fully completed by the Department Head, Nurse Manager, or Nurse-in-Charge or their representative and submitted within 72 hours to the appropriate administrator. Any delay in these reporting procedures must be accompanied by a written explanation for the lateness, and an endorsement by the appropriate administrator.

DEPT. HEAD/NURSE MANAGER/NURSE-IN-CHARGE/EMPLOYEE

1. COMPLETE LIGHT GREEN COPY of form and forward immediately to Safety Director.
2. PUMPKIN COLORED SHEET to the Head Nurse or Dept. Head. (EXCEPTION: SEE VISITOR NOT ON NURSING DIVISION)
3. COMPLETE THE FOLLOWING ON WHITE SHEET—FRONT AND BACK
    a. Description of Incident (what happened) by Patient, Visitor or Employee.
    b. What was accomplished prior to this incident to prevent its occurrence?
    c. Obtain Report and Signature of physician involved in treating patient.
    FORWARD FORM TO RESPONSIBLE ADMINISTRATOR.

ADMINISTRATOR

When the Administrator has completed his/her investigation, forward form to Safety Director.

VISITOR NOT ON NURSING DIVISION

Employee witnessing, involved in or informed of an incident should take visitor to the Emergency Room. IF VISITOR REFUSES TO GO TO EMERGENCY ROOM, employee should obtain **PATIENT–VISITOR INCIDENT REPORT** form from nearest department (DO NOT DETACH), obtain as much information as possible on LIGHT GREEN SHEET and forward ENTIRE FORM immediately to Safety Director.

EMERGENCY ROOM

With these incidents, the staff should complete the following portions of the **PATIENT–VISITOR INCIDENT REPORT** form:

1. LIGHT GREEN SHEET of Patient-Visitor Incident Report.
2. WHITE SHEET—Description of incident (what happened) by Visitor. Obtain physician's report and signature.

Contact Administrator on Duty for verification of charging or discount. Attach copy of Emergency Room note and forward completed form to Safety Director.

THIS DATA IS PROVIDED FOR THE LEGAL COUNSEL OF THE DIRECTORS OF URBAN HOSPITAL IN THE EVENT OF POSSIBLE LITIGATION AND IS TO BE CONSIDERED CONFIDENTIAL AND PRIVILEGED INFORMATION.

MEMO 7   Attachment A

GREEN COPY
INCIDENT REPORT / Patient-Visitor

THIS DATA IS PROVIDED FOR THE LEGAL COUNSEL OF THE DIRECTORS OF URBAN HOSPITAL IN THE EVENT OF POSSIBLE LITIGATION AND IS TO BE CONSIDERED CONFIDENTIAL AND PRIVILEGED INFORMATION.

USE ADDRESSOGRAPH IF PATIENT

REPORT NO.	PERSON INVOLVED			Age 72	Date of incident	Time (Military) 1000	Tele. No. 3092
	Goldman (Last Name)	Lucille (First Name)	F (M.I.)	Sex F	Date reported	Exact location of incident 4W	

PATIENT ☐

Rm. No. 410	Reason for hospitalization (Diagnosis) hypertension	Attending Physician Dr. Smith
Mental condition of patient before incident: Normal ☒ Senile ☐ Disoriented ☐ Sedated ☐ Other_____		Bedrails: Up ☒ Down ☐ Restraints: Yes ☐ No ☐ Activity Orders: Restraints ☐ Bed Rest: ☒ Up priviliges with assistance ☐ Without assistance ☐

Brief description of incident:
Patient given extra dose of Aldomet 250 mg at 1000

VISITOR ☐
OTHER ☐

By whom employed _____	Occupation _____
Home address_____	Home Phone_____
Nature of incident	Reason in Hospital

ACCIDENT FACTS

Name, Address, Tele. No. of Witnesses, if any:

Was patient seen by physician: Yes ☒   No ☐   Not indicated ☐   Physician's Name
Time called __1005__ a.m.   Time arrived __NA__ a.m.
                           p.m.   p.m.
Was treatment initiated by physician: Yes ☒   No ☐   X-Rays: Yes ☐   No ☒

I DO NOT WISH TO BE EXAMINED BY A PHYSICIAN: Signed: _____

MEMO 7 Attachment B

INCIDENT REPORT / Patient-Visitor

THIS DATA IS PROVIDED FOR THE LEGAL COUNSEL OF THE DIRECTORS OF URBAN HOSPITAL IN THE EVENT OF POSSIBLE LITIGATION AND IS TO BE CONSIDERED CONFIDENTIAL AND PRIVILEGED INFORMATION.

USE ADDRESSOGRAPH IF PATIENT

REPORT NO.	PERSON INVOLVED			Age	Date of incident	Time (Military)	Tele. No.
	(Last Name)	(First Name)	(M.I.)	Sex	Date Reported	Exact location of incident	

PATIENT ☐

Rm. No.	Reason for hospitalization (Diagnosis)	Attending Physician

Mental condition of patient before incident:
Normal ☐   Senile ☐
Disoriented ☐   Sedated ☐
Other _____

Bedrails: Up ☐   Down ☐   Restraints: Yes ☐   No ☐
Activity Orders: Restraints ☐   Bed Rest: ☐
Up priviliges with assistance ☐   Without assistance ☐

Brief description of incident:

VISITOR ☐
OTHER ☐

By whom employed _____   Occupation _____
Home address _____   Home Phone _____
Nature of incident   Reason in Hospital

ACCIDENT FACTS

Name, Address, Tele. No. of Witnesses, if any:

Was patient seen by physician: Yes ☐   No ☐   Not indicated ☐ | Physician's Name
Time called _____ a.m./p.m.   Time arrived _____ a.m./p.m.
Was treatment initiated by physician: Yes ☐   No ☐   X-Rays: Yes ☐   No ☐
I DO NOT WISH TO BE EXAMINED BY A PHYSICIAN: Signed: _____

DESCRIPTION OF INCIDENT

State what you saw and/or what you were told. Give names and addresses of all individuals who provided information concerning this incident.

Mrs. Goldman was given a dose of 250 mg of Aldomet by Kim Kas LPN at approximately 0945. Dose not charted and was repeated by myself (Nancy Noonan RN) at 1000 while I was passing medications. Dr Smith was notified immediately.

Date of Report	Signature of person preparing report	*Nancy Noonan*

PROPERTY DAMAGE ☐
MISSING ARTICLE ☐

Owner of Property	If theft, what day	and hour	last seen
Home Address NA		Tele. No.	

Nature and Extent of Damage or Loss

Estimated replacement or repair cost: $

MEMO 8 (see Attachments A and B)

INCIDENT REPORT / Patient-Visitor

NOTICE: Consider whether Security should be called immediately, especially in cases of theft or disappearance of patient or employee valuables, hospital property, including narcotics, keys to offices or narcotics cabinets, automobiles and contents thereof, as well as vandalism, fights, or threats of physical harm.

"An incident is any happening which is not consistent with the routine operation of the hospital or the routine care of a particular patient. It may be an accident or a situation which might result in an accident."

INSTRUCTIONS FOR USE

TIMELY SUBMISSION

The last part of form (White Sheet—front and back) must be fully completed by the Department Head, Nurse Manager, or Nurse-in-Charge or their representative and submitted within 72 hours to the appropriate administrator. Any delay in these reporting procedures must be accompanied by a written explanation for the lateness, and an endorsement by the appropriate administrator.

DEPT. HEAD/NURSE MANAGER/NURSE-IN-CHARGE/EMPLOYEE
1. COMPLETE LIGHT GREEN COPY of form and forward immediately to Safety Director.
2. PUMPKIN COLORED SHEET to the Head Nurse or Dept. Head. (EXCEPTION: SEE VISITOR NOT ON NURSING DIVISION)
3. COMPLETE THE FOLLOWING ON WHITE SHEET—FRONT AND BACK
    a. Description of Incident (what happened) by Patient, Visitor or Employee.
    b. What was accomplished prior to this incident to prevent its occurrence?
    c. Obtain Report and Signature of physician involved in treating patient.
    FORWARD FORM TO RESPONSIBLE ADMINISTRATOR.

ADMINISTRATOR

When the Administrator has completed his/her investigation, forward form to Safety Director.

VISITOR NOT ON NURSING DIVISION

Employee witnessing, involved in or informed of an incident should take visitor to the Emergency Room. **IF VISITOR REFUSES TO GO TO EMERGENCY ROOM**, employee should obtain **PATIENT–VISITOR INCIDENT REPORT** form from nearest department (**DO NOT DETACH**), obtain as much information as possible on **LIGHT GREEN SHEET** and forward **ENTIRE FORM** immediately to Safety Director.

EMERGENCY ROOM

With these incidents, the staff should complete the following portions of the PATIENT–VISITOR INCIDENT REPORT form:

1. LIGHT GREEN SHEET of Patient-Visitor Incident Report.
2. WHITE SHEET—Description of incident (what happened) by Visitor. Obtain physician's report and signature.

Contact Administrator on Duty for verification of charging or discount. Attach copy of Emergency Room note and forward completed form to Safety Director.

THIS DATA IS PROVIDED FOR THE LEGAL COUNSEL OF THE DIRECTORS OF URBAN HOSPITAL IN THE EVENT OF POSSIBLE LITIGATION AND IS TO BE CONSIDERED CONFIDENTIAL AND PRIVILEGED INFORMATION.

MEMO 8   Attachment A

GREEN COPY
INCIDENT REPORT / Patient-Visitor

THIS DATA IS PROVIDED FOR THE LEGAL COUNSEL OF THE DIRECTORS OF URBAN HOSPITAL IN THE EVENT OF POSSIBLE LITIGATION AND IS TO BE CONSIDERED CONFIDENTIAL AND PRIVILEGED INFORMATION.

USE ADDRESSOGRAPH IF PATIENT

REPORT NO.	PERSON INVOLVED			Age 62	Date of incident	Time (Military) 1925	Tele. No. 3092
	Brown	Cornelius	F	Sex M	Date reported	Exact location of incident 2 South	
	(Last Name)	(First Name)	(M.I.)				

PATIENT ☐

Rm. No. 406   Reason for hospitalization (Diagnosis): diabetes   Attending Physician: Dr. Green

Mental condition of patient before incident:
Normal ☒   Senile ☐
Disoriented ☐   Sedated ☐
Other _____

Bedrails: Up ☐ Down ☒   Restraints: Yes ☐ No ☒
Activity Orders: Restraints ☐   Bed Rest: ☐
Up privileges with assistance ☐   Without assistance ☒

Brief description of incident:
PT fell while getting out of bed

VISITOR ☐
OTHER ☐

By whom employed _____   Occupation _____
Home address _____   Home Phone _____
Nature of incident   Reason in Hospital

ACCIDENT FACTS

Name, Address, Tele. No. of Witnesses, if any:

Was patient seen by physician: Yes ☒   No ☐   Not indicated ☐   Physician's Name
Time called __1930__ a.m. / (p.m.)   Time arrived __1945__ a.m. / (p.m.)
Was treatment initiated by physician: Yes ☐   No ☒   X-Rays: Yes ☐   No ☒
I DO NOT WISH TO BE EXAMINED BY A PHYSICIAN: Signed: __N/A__

MEMO 8 Attachment B

INCIDENT REPORT / Patient-Visitor

THIS DATA IS PROVIDED FOR THE LEGAL COUNSEL OF THE DIRECTORS OF URBAN HOSPITAL IN THE EVENT OF POSSIBLE LITIGATION AND IS TO BE CONSIDERED CONFIDENTIAL AND PRIVILEGED INFORMATION.

USE ADDRESSOGRAPH IF PATIENT

REPORT NO.	PERSON INVOLVED (Last Name) (First Name) (M.I.)	Age	Date of incident	Time (Military)	Tele. No.
		Sex	Date Reported	Exact location of incident	

PATIENT ☐

Rm. No.	Reason for hospitalization (Diagnosis)	Attending Physician

Mental condition of patient before incident:
Normal ☐   Senile ☐
Disoriented ☐   Sedated ☐
Other_____

Bedrails: Up ☐ Down ☐   Restraints: Yes ☐ No ☐
Activity Orders: Restraints ☐   Bed Rest: ☐
Up privileges with assistance ☐   Without assistance ☐

Brief description of incident:

VISITOR ☐
OTHER ☐

By whom employed _____   Occupation _____
Home address _____   Home Phone _____
Nature of incident   Reason in Hospital

ACCIDENT FACTS

Name, Address, Tele. No. of Witnesses, if any:

Was patient seen by physician: Yes ☐ No ☐ Not indicated ☐   Physician's Name
Time called _____ a.m./p.m.   Time arrived _____ a.m./p.m.
Was treatment initiated by physician: Yes ☐ No ☐   X-Rays: Yes ☐ No ☐
I DO NOT WISH TO BE EXAMINED BY A PHYSICIAN: Signed: _____

DESCRIPTION OF INCIDENT

State what you saw and/or what you were told. Give names and addresses of all individuals who provided information concerning this incident.

Mr. Brown called out to desk and told the unit secretary that he had fallen and needed help. I (Pam Evans, RN) and Theresa Jones, LPN, found Mr. Brown in a sitting position on the floor near the bed. Call bell was in reach. Pre-fall vital signs at 1600 were 37-84-20 130/80. Vital signs after fall at 1930 were 36.5-90-24 134/86. No contusions or abrasions noted. No c/o pain.

Date of Report	Signature of person preparing report *Pam Evans, RN*

PROPERTY DAMAGE ☐
MISSING ARTICLE ☐

Owner of Property	If theft, what day	and hour	last seen
Home Address NA			Tele. No.
Nature and Extent of Damage or Loss			
			Estimated replacement or repair cost: $

MEMO 9

> Janet,
>
> We are having a great deal of conflict on the 11–7 shift over assigned duties of LPNs and RNs. I feel that I need help in dealing with the problem before it gets out of hand. Perhaps we could set up some type of discussion. Let me know what you think soon!
>
> Beverly
(Asst. Nurse Manager)

MEMO 10

> Janet,
>
> Bev Myers in Room 06 wants to speak with you immediately about the nurse who cared for her last evening. She's tremendously angry but says she will not speak to anyone but the nurse manager.
>
> Pat
(Staff Nurse, 7–3)

MEMO 11

> Janet,
>
> I need off next weekend if at all possible. My son is graduating from high school, and it means so much to have the whole family participate in the weekend activities. Please let me know as soon as possible.
>
> Marty Hays
(Nurse Aide, 7–3)

MEMO 12

Janet,

Every evening this week something has been missing from the division, ranging from a snack in the fridge (with a name on it) to money from a wallet. While at first I assumed things were being misplaced, I now suspect (as many others do) these items are being stolen.

The staff is becoming very annoyed. I've never dealt with such a situation. Could I talk to you about it soon? Thanks.

Beverly
(Asst. Nurse Manager)

MEMO 13

HOSPITAL NURSING SERVICE

TO: Janet Baccard
Nurse Manager
FROM: Patty Lenahan
Dietary
DATE: June 17
RE: Patient Food Trays

We need to meet with you to discuss the patient food trays arriving on your division. We're checking into patient complaints concerning accuracy, temperature, taste, etc. Please let me know of a convenient time for you.

MEMO 14 (see Attachment on page 114 and 115)

HOSPITAL NURSING SERVICE

TO: Janet Baccard
FROM: Linda Wilson
Product Evaluation Nurse
DATE: June 17
RE: IV Method Survey

Please complete the attached questionnaire and return to me by next Monday. Thanks for your help.

MEMO 14 Attachment

NURSING SURVEY
IV Methods Survey

1. Nursing Unit _____ Nurse Manager _____

2. No. of Beds _____

3. Average Daily Census _____

4. Average Number of IVs (Primary lines only) _____

5. How often are IV Tubings changed (check)

	Primary	Secondary (piggyback line)
Q 24 hour	_____	_____
Q 48 hour	_____	_____
With every bag or bottle	_____	_____

6. Are you using extension sets and stopcocks routinely? Why?
   _____
   _____

7. How many IVs need to be restarted per day (24 hour) on the average, due to infiltrates or other *problems* (not normal set changes)?
   _____

8. List the IV set numbers which you use on a *routine* basis only.

   _____   _____   _____

   _____   _____   _____

9. Are volume limiting sets (Buretrol or Soluset sets) being used? _____

   How often? _____

   Why? _____

10. When patients go to surgery *or* come from the emergency room, do they come to your floor on the type of IV tubing that you use routinely or on different sets (i.e., mini drips, "Y" tubing, etc.)?
    _____
    _____

11. Are filters being added to any IVs on your floor?
    _____

    Under what circumstances? _____

MEMO 14 Attachment, *continued*

12. Is piggybacking (adding medications) done by:

    _____ 1. Using Buretrol or Soluset.

    _____ 2. Using a *Short Add A Line* at the Upper Y site.

    _____ 3. Using a *long line* plugged into the *bottom* Y site or stopcock.

13. Are you presently using any infusion devices on your floor?
    _____

    What kind? _____

    Under what circumstances? _____

14. Which of the following problems associated with IV therapy most concern you? (Rank 1–6, worst problem being #1)

    _____ Infiltrates

    _____ Clogged Needles

    _____ Positional IVs

    _____ Runaway IVs

    _____ Kinked Tubing

    _____ Inaccurate Flow Rates

15. On an average, how many patients per day fall into the following IV categories:

    _____ IV rates below 5 cc/hour.

    _____ IV rates above 200 cc/hour.

16. In your opinion, what could be done to improve your current IV program?
    _____
    _____
    _____
    _____

Exercise 13–2 Recognizing Time Wasters

INTRODUCTION

Everyone wastes time; however, nurse managers who produce good results keep their wasted time to a minimum. Some of the most common time wasters for managers are interruptions such as unscheduled visitors, telephone calls, or crises; attendance at inefficient meetings; and other activities such as failure to delegate, routine tasks, and lack of a daily plan.

Drop-in visitors are a frequent source of wasted time due to the "open door" policy of many managers. The following are ways to reduce wasted time due to drop-in visitors.

1. Establish a "quiet time" when you have your door closed to visitors, preferably amounting to approximately 20 percent of your work day. Learn to say no to visitors during this time. Encourage staff to protect your time as well.
2. Schedule regular times with staff members to deal with their problems.
3. Rearrange your office to discourage drop-in visitors—angle your desk to 90 to 180 degrees from the door, remove unnecessary chairs.
4. Set time limits at the beginning of discussions.
5. Identify frequent visitors and schedule necessary meetings with them.

While these methods are sometimes difficult, they will result in more effective use of your time. Should you have an unexpected visitor, below is a list of key behaviors for dealing with drop-in visitors.

1. Listen carefully to the visitor's initial request.
2. Determine whether or not the visitor needs to talk to you now.
3. If it is not an emergency, briefly state that you are unable to meet now.
4. Set a time and place to meet with the person that is reasonably soon and mutually agreeable.
5. Avoid lengthy explanations and justifications for your actions.

Another common time waster for nurse managers is the telephone. Ways to reduce telephone time are as follows.

1. Develop a method to screen and delegate calls and appointments.
2. Examine your reasons for chit-chat. Keep irrelevant conversation to a minimum.
3. Have all materials at hand before calling.
4. Tell talkative callers that you have another call.
5. Establish a daily time period for receiving calls and tell everyone what it is.
6. If possible, suggest that routine information be provided by memo, electronic mail, or fax communication, rather than by telephone.

When you have someone who repeatedly calls you, refer to the key behaviors for dealing with telephone time wasters given below.

1. Plan your telephone call by organizing yourself before you call.
2. Start the call by identifying yourself and making a statement of purpose.
3. Keep "chit-chat" to a minimum.
4. Let the person know that you have limited time to talk now.
5. Get to the point quickly.
6. Terminate the conversation when your agenda is complete, without fearing that the other person will be offended.

EXERCISE

TYPE OF ACTIVITY: SMALL GROUP

Instructions

1. Divide into groups as directed.
2. Review the key behaviors for dealing with a drop-in visitor listed in the introduction. Then, take turns role-playing a situation where you are a manager with a drop-in visitor. The third person in each group should be an observer.
3. After each has had an opportunity to manage a drop-in visitor, review the key behaviors for dealing with telephone time wasters listed in the introduction. Then, role-play a situation in which you are a manager with a telephone caller. The participants should sit back to back to simulate an actual telephone call in which the caller cannot be seen.

Discussion

1. Which of the time wasters listed above are you more likely to find yourself doing?
2. What was the most difficult part of the role-play situation?
3. What are some ways to handle the other common time wasters?

Further Reading

Josephs, R. (1992). *How to Gain an Extra Hour Every Day: More than 500 Time Saving Tips*. New York: Penguin.

Exercise 13–3 Managing Your Professional Time

INTRODUCTION

Effective managers *plan* their time for maximum productivity. While unexpected circumstances do occur, effective managers have general guidelines for how their professional time is scheduled. Most managers use some type of scheduling calendar to plan their day.

EXERCISE

TYPE OF ACTIVITY: INDIVIDUAL

Instructions

Read through the scenario. Use the schedule provided (Figure 1 on page 120) to prioritize your day, based upon what you know about effective time management. Be prepared to justify your schedule.

Scenario

You are the manager of a home health care agency. The following activities are proposed for the following day:

1. Meeting with the community board of directors for the agency to discuss funding for the upcoming fiscal year. Time allotted: 2 hours.

2. Staff meeting with the home health aides to discuss the new intravenous pumps. Time allotted: 30 minutes.

3. Monthly case review with case managers. Time allotted: 1–2 hours, depending on the detail provided by the managers.

4. Lunch break with 30-minute walk for exercise. Time allotted: 1 hour.

5. Need to prepare for the meeting with the board. Time allotted: as much as possible.

6. Jeff, one of the nurses, has requested to discuss a personal problem with you. Time allotted: not specified.

7. Meet with the middle managers about new policies regarding the dress code and use of new cellular phones by the home health nurses. You are seeking the advice of these managers as you finalize revisions for the new policies. Time allotted: 1 hour.

8. Need to review your in-basket. Time allotted: 15 minutes–2 hours, depending upon whether you act on the items this day.

Discussion

1. What did you consider to be the priority tasks of the day?
2. When did you schedule these priority tasks? Why?

3. Did you have enough time to complete all of the activities? Why or why not?
4. What other strategies can be used to accomplish needed activities during one's workday?

Further Reading

Mackenzie, A. (1990). *The Time Trap*. New York: American Management Association.

FIGURE 1 Daily Schedule

Time	Appointments/ Scheduled Events	Expect to Accomplish
0800		
0900		
1000		
1100		
1200		
1300		
1400		
1500		
1600		
1700		
1800		

To Be Done Today:

MODULE 14 — DELEGATION

Exercise 14–1   Case Scenario: Use of Delegation

Exercise 14–2   Differentiating Responsibility, Accountability, and Authority

Exercise 14–1   Case Scenario: Use of Delegation

INTRODUCTION

Delegation is an increasingly important skill of nurses and nurse managers. As the demand for efficiency and effectiveness increases, nurse managers must delegate knowledgeably. Delegation is the process by which responsibility and authority for performing a task (function, activity, or decision) is transferred to another individual who accepts that authority and responsibility. One individual empowers another, creating a trusting relationship where both parties stand to gain. The delegator increases the amount of work that can be accomplished. The delegatee is provided an opportunity to grow professionally.

To effectively delegate, one must understand the difference between responsibility and accountability. Responsibility is an obligation to accomplish a task, whereas accountability is accepting ownership for the results or lack thereof. In delegation, responsibility is transferred while accountability is shared. It is important to remember that you can only delegate tasks for which you are responsible. The second component of delegation is authority. Along with responsibility, the delegator must transfer authority (the right to act). Without authority, the delegatee lacks the power to act. Giving the delegatee responsibility without authority to act is the primary source of failure in the delegation process.

The five steps of the delegation process are listed below:

1. Defining the task.
2. Determining to whom to delegate.
3. Providing clear communication about expectations regarding the task.
4. Reaching mutual agreement about the task at hand.
5. Monitoring and evaluating the results and providing feedback to the individual regarding their performance.

EXERCISE

Type of Activity: Small Group

Instructions

1. Divide into groups as instructed and select an observer-recorder.
2. Role-play the scenario as assigned.
3. The observer-recorder should note the group's activity.

Scenarios

1. You are the nurse manager of a postpartum mother-baby unit. As the need for early parenting education increases, you are aware that you need to delegate the education responsibilities to one of the staff. You believe that Eve, a staff nurse on 3–11, possesses the necessary skills and abilities to take on the job. You are not sure, however, whether Eve is interested. You make an appointment to discuss this possibility with her.

2. As the new staff nurse of a large oncology unit, you have been given the responsibility to develop a family support program for the families of terminal patients. You do not feel ready to accept this responsibility and have asked to talk to the nurse manager about your insecurities. You do recognize that this is a wonderful opportunity to learn more about management, but you need to talk about what delegation really means.

3. You are the nurse manager of a rapidly growing same-day surgery center. You feel overwhelmed with the work that is piling up and are looking for any way to unload some of it to the staff. A somewhat unreliable nurse has offered to take over the scheduling. You have some reservations, but feel you will know better about whether she is capable of accepting this accountability, responsibility, and authority after you talk to her.

Discussion

1. Report on the role-play. What did the nurse manager decide to do in each scenario?
2. How did the staff nurses feel when delegated to?
3. What caused difficulty with delegation?

Further Reading

American Nephrology Nurses' Association. (1992). Delegation of nursing tasks to licensed and unlicensed personnel: A guide for ESRD facilities. *ANNA Journal,* 19, 337–338.

Barter, M. & Furmedge, M. C. (1994). Unlicensed assistive delegation and supervision. *Journal of Nursing Administration,* 24(4), 36–40.

Exercise 14–2 Differentiating Responsibility, Accountability, and Authority

INTRODUCTION

As mentioned in the previous exercise, the process of delegation requires knowledge of the concepts of responsibility, accountability, and authority. It is often a breakdown in the transfer of one of these processes that causes delegation to be unsuccessful.

EXERCISE

Type of Activity: Small Group

Instructions

1. Read each of the following scenarios in class.
2. With group discussion, identify whether responsibility, accountability, and/or authority have been transferred, have not been transferred, or are currently shared.

Scenarios

1. You are the 3–11 charge nurse on a medical unit. Your nurse manager has instructed you to inform a 3–11 staff nurse that she has been terminated, based upon a progressive disciplinary plan that you have not been previously involved with.
2. You are the nurse manager of a busy labor and delivery unit. A staff RN has just expressed an interest in orienting the new unlicensed assistive personnel to the unit. You agree that it is a great idea.
3. Mr. Leland has been struggling to manage his evening wound care by himself at home. His health insurance will only cover daily RN visits, although you believe he warrants twice daily dressing changes. As the case manager for Mr. Leland, you instruct the home health aide to report to you any deterioration of the wound when she sees him in the evenings.
4. The staff has requested a self-governance structure. Most specifically, they have expressed an interest in self-scheduling. You agree to let them cover the next month's schedule, but state, "If anyone is unhappy with the final schedule, don't tell me about it."

Discussion

1. In each of the scenarios, what is happening in terms of responsibility, accountability, and authority?
2. In each of the scenarios, what can be done to rectify the situation, if needed?
3. As a manager, is it easier to transfer responsibility, accountability, or authority? Why?

Further Reading

American Association of Critical Care Nurses. (1990). *Delegation of Nursing and Non-Nursing Activities in Critical Care: A Framework for Decision Making.* Irvine, CA: American Association of Critical Care Nurses.

Manthey, M. (1990). Trust: Essential for delegation. *Nursing Management,* 21(1), 28–31.

MODULE 15: STRESS ASSESSMENT AND REDUCTION

Exercise 15–1   Reducing Organizational Stress

Exercise 15–1   Reducing Organizational Stress

INTRODUCTION

Stress is a daily part of our lives. However, stress will often impede employees' productivity and cause an overall decrease in the function of the organization. The source of stress can be organizational or personal; often, it is a combination of both. It is the nurse manager's responsibility to recognize the potential organizational stressors for employees and identify ways to manage the stress response felt by employees. It is also important to recognize when an employee's personal stressors are causing the employee to be increasingly stressed at work. An astute manager recognizes when staff are becoming overly stressed and helps employees find ways to reduce and manage their stress experience.

EXERCISE

TYPE OF ACTIVITY: LARGE GROUP

Instructions

1. Read the following scenario.
2. Be prepared to discuss the scenario and answer the following questions.

Scenario

Bev is a 28-year-old nurse working 11–7 on your unit. She has been employed by your organization for six years and requested to remain on night shift, despite having the necessary seniority to work days or evenings. The primary reason for this request is the recent birth of her first child, now four months old. She and her husband have been working different shifts to provide childcare for their son.

Bev is an excellent nurse, but since returning to work, her effectiveness has decreased. She often works more than 40 hours per week. Her number of medication errors has increased, and she has had some interpersonal conflicts with one of the unlicensed assistive personnel. Bev appears tired and is often short-tempered with all of the staff.

During her maternity leave of absence, the unit underwent some changes. The unit documentation became computerized and the unit adopted new intravenous medication pumps. These changes have been difficult for Bev because she missed both orientation sessions. You suspect that these changes are part of her productivity problem. You suspect, however, that it is not the entire cause. As her nurse manager, you schedule a meeting with Bev to discuss her recent performance.

Discussion

1. Identify the organizational and personal stressors of the employee.
2. How is Bev's stress being manifested at work?
3. What are your suggested strategies for stress management and stress reduction?

Further Reading

Lazarus, R. & Folkman, S. (1984). *Stress, Appraisal and Coping.* New York: Springer.

SECTION VI
MANAGING HUMAN RESOURCES

MODULE 16 INTERVIEWING

Exercise 16–1 Developing a Structured Interview Guide
Exercise 16–2 Effective Interviewing

MODULE 17 EVALUATIONS

Exercise 17–1 How to Write a Critical Incident

MODULE 18 STAFF DEVELOPMENT

Exercise 18–1 Coaching as a Staff Development Strategy
Exercise 18–2 Providing a Formal Performance Appraisal
Exercise 18–3 Multicultural Issues in Staff Development

MODULE 19 MOTIVATION

Exercise 19–1 Personal Motivation: Use of Goal Setting
Exercise 19–2 Employee Motivation

MODULE 16 INTERVIEWING

Exercise 16–1   Developing a Structured Interview Guide

Exercise 16–2   Effective Interviewing

■ Exercise 16–1 Developing a Structured Interview Guide

INTRODUCTION

The interview remains the single most common technique used in the hiring process. The interview is an information-seeking meeting between an individual applying for a position and a member of an organization doing the hiring. In the interview, the interviewer is trying to evaluate information that has been gathered from the application form, interview, and from tests. The applicant is trying to gather information about the job and the hospital. Selection interviewing may be conducted by a personnel department, by nurse managers, or both.

To be an effective interviewer, you must learn to gather relevant and appropriate data. The basic interviewing skills needed are as follows.

1. *Planning the interview* Become familiar with the application form, the job requirements, and areas to be covered in the interview. Plan and organize questions pertinent to the job and the applicant. Prepare for the interview in an environment free from interruption.

2. *Presenting oneself* The interviewer makes an impression on the applicant, both as an individual and as a representative of the organization. This includes the applicant's impression of the interviewer's tone of voice, eye contact, personal appearance and grooming, posture, and gestures.

3. *Responding to the applicant* Maintain concern for the applicant's feelings while controlling the interview. React appropriately to the

applicant's comments, questions, and nonverbal behaviors. Convey interest in the applicant, encourage an atmosphere of warmth and trust, and make use of encouragement and praise.

4. *Getting information*   Use appropriate questioning techniques to elicit relevant information. Probe incomplete answers and problem areas while maintaining an atmosphere of trust.

5. *Giving information*   Communicate appropriate and accurate information about the institution and available jobs for which the applicant would qualify. Answer any questions the applicant may have.

6. *Information processing*   Integrate and analyze interview information for a final placement decision. Identify personal characteristics and judge them in the context of the job requirements. The interviewer needs skill in assimilating, remembering, and integrating all information that is relevant to the final evaluation.

Although the interview is widely used, research suggests that the interview is not highly accurate and can easily become a very subjective selection tool. Selection errors in interviewing can be caused by the interviewer or the process. Not only are interviewers likely to inject their personal prejudices or first impressions into the selection decision, but they also may not be using effective or relevant questions to gain information from the candidate. Furthermore, the most accurate selection decisions are those made by choosing the best of several applicants rather than a yes or no decision made for each applicant as he or she is interviewed.

Structuring the interview increases its effectiveness by helping reduce the interviewer's bias, supplying relevant and effective information, and giving the interviewer the same basic information on all candidates. Furthermore, a structured interview guide can be designed for easy note-taking. Notes will increase interviewer recall of information about each applicant and can be stored to later justify the selection decision, if required under an equal employment opportunity complaint.

EXERCISE

Type of Activity: Small Group

Instructions

1. Read the job description for the Unit Nursing Director (Figure 1 on page 134).

2. For each meaningful category of tasks, individually outline what you believe to be the abilities required. Consider mental abilities (spelling, basic math, decision making), psychomotor skills, job knowledge, and personality traits. Do not use easily trainable skills, abilities, and knowledge as selection criteria. In addition, avoid

questions that might be discriminatory. Table 1 on page 135 outlines appropriate and inappropriate questions.

3. Divide into small groups as instructed.
4. Together decide on and outline ways to measure, in an interview, the skills and abilities required for the job. Consult the types of questioning techniques listed in Table 2 on page 137.
5. Develop a structured interview guide. Be prepared to present your guide to the class.

Discussion

1. How does the structured interview guide improve selection?
2. What are some of the questions and issues to avoid in the interview?
3. How can this experience help you in interviewing for a job or in interviewing others for a job?

Further Reading

Gatewood, R. D. & Feild, H. S. (1990). *Human Resource Selection*. Chicago, IL: Dryden.

Secatore, J. A. & Stengrerics, S. S. (1994). In search of a perfect match. In R. Spitzer-Lehmann (Ed.), *Nursing Management Desk Reference: Concepts, Skills, and Strategies*. Philadelphia: W. B. Saunders, pp. 159–177.

Tibles, L. R. (1993). The structured interview: An effective strategy for hiring. *Journal of Nursing Administration*, 23(10), 42–46.

FIGURE 1 Sample Job Description

DEPARTMENT OF NURSING SERVICE
2nd Floor Medical Unit, 3rd Floor Surgical Unit, Skilled Nursing Facility
JOB DESCRIPTION, UNIT NURSING DIRECTOR

BASIC FUNCTIONS

Functions as the Director of Nursing Care for the unit(s) and is accountable for the quality of nursing care provided in areas of responsibility.

JOB RESPONSIBILITIES

1. *Supports the vision and mission* of LMH and the philosophy of the Department of Nursing.
2. On an ongoing basis, through interaction with Nursing Department Directors, members of hospital administration, and community, *assesses the quality of nursing care* provided via a variety of avenues, i.e. personal interaction, patient questionnaires, formal and informal data collections.
3. On an ongoing basis, using a multitude of data sources, i.e. literature, educational opportunities, *assesses the trend in health care/nursing care* and the impact of these trends on the Nursing Department of LMH.
4. Through an analysis of trends in health care/nursing care and community needs/perception of nursing care at LMH; *plans for departmental growth and development* and maintenance of current programs, i.e., professional development; nursing care delivery model.
5. Working with Nursing Department Directors, other hospital departments/administration and the members of the community, *implements appropriate strategies/processes* to maintain a competent nursing staff skilled in the provision of care and resources to meet the nursing care needs of the members of the Lawrence community and service area.
6. In conjunction with other members of the Department of Nursing, *develops a budget* annually which is compatible with institutional resources and *provides the means* necessary to maintain competent staff, *provides quality patient care* and *meets the needs of the medical staff. Monitors* units' monthly expenditures.
7. *Organizes and conducts* unit meetings for communication and education.
8. *Responsible* for unit monthly staffing schedule.
9. *Collaborates with Nursing Supervisors and Staffing Coordinator* to adjust daily staffing needs determined by patient acuity/unit needs.
10. *Models CQI behavior* in problem solving and decision making process.
11. *Ability to identify* personal learning needs and utilize time management skills.
12. *Assists in developing* unit level QA Plan. Develops monitors for QA Plan and participates in audits. Actively participates in the Nursing QA Committee.
13. *Actively co-chairs* a hospital committee.
14. *Participates* in NSAC.
15. *Assists* in unit staff orientation.
16. *Evaluates staff members* on a continuing basis providing constructive feedback for professional and personal growth.
17. *Demonstrates* appropriate interpersonal skills.

KNOWLEDGE, SKILLS AND ABILITIES

1. Current license to practice as a Registered Nurse in the State of Kansas, and hold a BSN degree.
2. Manual dexterity sufficient to execute written communications.
3. Written and verbal communication skills to interact effectively with a variety of personalities and departments.
4. Sufficient ambulatory capabilities to make rounds on Nursing Units entering nurses station, patient rooms and treatment areas.
5. High level time management skills with ability to perform a multitude of tasks simultaneously and set priorities/time frame for major project completion.

WORKING CONDITIONS

Office environment; low to moderate noise level; frequent interruptions; demanding task requirements; risk of exposure to communicable diseases.

This job description in no way states or implies that these are the only activities to be performed by the employee occupying this position. Employees will be required to follow any other job-related instructions and to perform any other job-related responsibilities requested by their Vice-President/Nursing.

Source: Obtained with permission from Lawrence Memorial Hospital, Lawrence, Kansas.

TABLE 1	PREEMPLOYMENT QUESTIONS	
	APPROPRIATE TO ASK	**INAPPROPRIATE TO ASK**
NAME	■ Applicant's name. Whether applicant has school or work records under a different name.	■ Questions about any name or title that indicate race, color, religion, sex, national origin or ancestry. Questions about father's surname or mother's maiden name.
ADDRESS	■ Questions concerning place and length of current and previous addresses.	■ Any specific probes into foreign addresses that would indicate national origin.
AGE	■ Requiring proof of age by birth certificate after hiring. Can ask if applicant is between 18 and 70.	■ Requiring birth certificate or baptismal record *before* hiring.
BIRTHPLACE OR NATIONAL ORIGIN		■ Any questions about place of birth of applicant or place of birth of parents, grandparents or spouse. ■ Any other questions (direct or indirect) about applicant's national origin.
RACE OR COLOR	■ Can request *after* employment as affirmative action data.	■ Any inquiry that would indicate race or color.
SEX		■ Any question on an application blank that would indicate sex.
RELIGION		■ Any questions to indicate applicant's religious denomination or beliefs. ■ Request a recommendation or reference from the applicant's religious denomination.
CITIZENSHIP	■ Question about whether the applicant is a U.S. citizen; if not, whether the applicant intends to become one. ■ Question if applicant's U.S. residence is legal and require proof of citizenship after being hired.	■ Questions of whether the applicant, his/her parents, or spouse are native born or naturalized. ■ Require proof of citizenship before hiring.
PHOTOGRAPHS	■ May require after hiring for identification purposes only.	■ Request photograph *before* hiring.
EDUCATION	■ Questions concerning any academic, professional, or vocational schools attended. ■ Inquiry into language skills, such as reading and writing of foreign language.	■ Questions asking specifically the nationality, racial or religious affiliation of any school attended. ■ Inquiries into the applicant's mother tongue or how any foreign language ability was acquired (unless it is necessary for the job).

(continued)

TABLE 1   PREEMPLOYMENT QUESTIONS, *continued*

	APPROPRIATE TO ASK	INAPPROPRIATE TO ASK
RELATIVES	■ Ask for the name, relationship and address of a person to be notified in case of an emergency.	■ Any unlawful inquiry about a relative or residence mate(s) as specified in this list.
CHILDREN		■ Questions about the number and ages of children or information on childcare arrangements.
TRANSPORTATION		■ Inquiries about transportation to or from work (unless car necessary for job).
ORGANIZATION	■ Questions about organization memberships and any offices that might be held.	■ Questions about any organization an applicant belongs to which may indicate the race, color, religion, sex, national origin or age or ancestry of its members.
PHYSICAL CONDITION/DISABILITIES	■ Questions about meeting the job requirements, with or without some accommodation.	■ Questions about general medical condition, state of health, specific diseases, or nature/severity of disability.
MILITARY SERVICE	■ Questions about services rendered in armed forces, the rank attained, and which branch of service. ■ Require military discharge certificate *after* being hired.	■ Questions about military service in any armed forces other than the U.S. ■ Request of military service records before hiring.
WORK SCHEDULE	■ Questions about the applicant's willingness to work required work schedule.	■ Ask applicant's willingness to work any particular religious holiday.
REFERENCES	■ Ask for general and work references not relating to race, color, religion, sex, national origin or ancestry, age, disability	■ Request references specifically from clergymen (as specified above) or any other persons who might reflect race, color, religion, sex, national origin or ancestry of applicant, age, disability.
FINANCIAL		■ Questions about banking, credit rating, outstanding loans, bankruptcy, or having wages garnished.
OTHER QUALIFICATIONS	■ Any question that has direct reflection on the job to be applied for.	■ Any non-job-related inquiry that may present information permitting unlawful discrimination. Questions about arrests or convictions (unless necessary for job, such as security clearance).

Source: Ohio Civil Rights Commission and United States Equal Employment Opportunities Commission (1992). Americans with Disabilities Act (EEOC Publication #M-1A). Washington, DC: US Government Printing Office.

TABLE 2 — TIPS FOR INTERVIEWERS

While you are watching job candidates for cues about how they feel, they will be watching you. Applicants will try to find out whether you are interested in what they are saying, whether you approve or disapprove of them, and whether you wish the interview were over already! The cues that the person reads in your behavior can serve to make the candidate more relaxed and allow open communication, or they can make the interviewee more tense and nervous; this causes the applicant to become defensive. Since your responsiveness has a strong influence on the success of the interview, you must be constantly aware of the impressions you are giving to the applicant. These impressions can be influenced by both your verbal and nonverbal communication.

NONVERBAL SIGNS

1. Head nodding indicates that you understand and agree with the candidate.
2. Smiling or laughing when appropriate indicates that you enjoy the person's company and are listening with interest.
3. Direct eye contact shows interest.
4. A relaxed manner (leaning back in your chair) shows you are willing to listen.
5. Leaning forward indicates interest—that you want the person to continue talking.
6. Silence can indicate a desire for the candidate to continue.
7. Glancing at your watch or around the room indicates boredom, eagerness for the interview to be concluded.
8. Shuffling through papers indicates lack of interest.
9. Constantly referring to a list of questions or the application form indicates lack of confidence in your skill as an interviewer.
10. Concentrating more on note-taking than watching the applicant shows you are more interested in filling out a form than getting to know the applicant.
11. Showing expressions of surprise, shock, or annoyance indicates you are judging the person and makes the applicant regret his or her honesty; this can destroy rapport.

VERBAL RESPONSES

1. Encouraging comments (good, I see, I understand) indicate you are listening, are interested, and agree with what the applicant says, as long as they are inserted at appropriate times and seem a reflection of genuine interest. An absent-minded comment is far worse than none at all.
2. Supportive remarks and praise ("It's wonderful that you were able to accomplish so much in such a short time") indicate that you recognize his or her achievements. They encourage the person to say more.
3. Playing down negative information ("I certainly understand how that could have happened under those circumstances") reassures the candidate that you aren't judging the person unfairly, and it also increases the possibility that he or she will reveal additional information that the person might otherwise have tried to hide.
4. Restatement of the candidate's thought ("you were interested in that job because of the advancement potential") indicates that you follow what has been said, and it may encourage further elaboration. Take care in using this technique, however, so that you don't become a "parrot" or try to put words in the applicant's mouth.
5. Interrupting or changing the subject abruptly indicates a lack of interest in what has been said and a lack of courtesy in turning the conversation to a more relevant topic. This can be particularly damaging if you stop the candidate before the person has reached the point of the comment. For example:

   Applicant: There was only one thing that made me stick with that job so long . . .
   Interviewer: Yes, first jobs are frequently that way. Now, tell me more about your organizing the hospital softball team.

   This type of switch makes the applicant feel a little foolish and, at the same time, resentful toward you for your lack of interest in the story.

Exercise 16–2 Effective Interviewing

INTRODUCTION

An effective interviewer must learn to obtain the needed information in a manner that is effective, while maintaining an overall pleasant tenor of the meeting. Interviews typically will last from 1 to 1½ hours, and should include an opening, an information gathering and giving phase, and a closing.

The key behaviors for effective interviewing are listed in Table 1. Additional interviewing tips are provided in Table 2 of Exercise 16-1. Effective interviewing requires that the interviewer be prepared for the interview by reading the résumé and preparing the setting for a relaxed, yet professional meeting. Opening the interview includes starting the interview on time, establishing rapport with the interviewee, and providing an outline of the interview process. Early in the interview, the interviewer should try to avoid making hasty first impressions. Next, the interviewer should follow a structured interview guide to reduce bias, to ensure that relevant information is obtained, and to provide consistency among interviews. At the completion of the interviewer's questioning, the interviewee should be encouraged to ask questions if the interviewer feels that this individual is a promising candidate for the position.

EXERCISE

TYPE OF ACTIVITY: SMALL GROUP

Instructions

1. Review Table 1, Key Behaviors for Interviewing.
2. Divide into groups of three and assign the roles of Interviewer, Interviewee, and Observer.
3. Assume that the Interviewer is the nurse manager of a skilled nursing facility, and the Interviewee is a new graduate seeking a first job.
4. Practice each role for 10 minutes, switching roles after the Observer has provided feedback for each role-play.

TABLE 1	KEY BEHAVIORS FOR INTERVIEWING
OPENING THE INTERVIEW	1. Give a friendly greeting—relax and smile. 2. Talk about yourself to put the person at ease. 3. Tell the person what you will cover during the interview.
OBTAINING INFORMATION	1. Probe for unfavorable information. 2. Seek information to substantiate or refute your initial impressions. 3. Avoid leading questions. 4. Ask only one question at a time. 5. Use words such as how, what, why, and tell me. 6. Adhere to your structured interview guide.
PROVIDING INFORMATION	1. Answer direct questions. 2. Provide detailed job information only if you believe the applicant to be a serious candidate.
CLOSING THE INTERVIEW	1. Summarize the applicant's strengths and weaknesses. 2. Ask if he or she has anything to add. 3. Tell him or her the next steps in the employment process. 4. Thank the applicant. 5. Complete interview notes after the interview.

Discussion

1. What are your strengths and weaknesses as an interviewer?
2. How does your evaluation compare with that of others in your group?

Further Reading

Arvey, R. D. & Faley, R. H. (1988). *Fairness in Selecting Employees.* Reading, MA: Addison-Wesley.

MODULE 17 EVALUATIONS

Exercise 17–1   How to Write a Critical Incident

■ Exercise 17-1 How to Write a Critical Incident

INTRODUCTION

Nurse managers need continuous documentation of an employee's performance to provide the basis for an accurate performance appraisal. An excellent strategy for keeping such documentation is to encourage the use of the critical incident technique. In this technique, peers and managers document events using an objective, behaviorally-based format. Subjective statements involving opinion are avoided. Key behaviors for writing critical incidents are listed below.

1. Identify what constitutes the critical performance of each job description.
2. Provide a succinct, behaviorally-based objective description of the performance.
3. Write the outcome of the performance.
4. If appropriate, state what was done to correct the performance deficiency or to reward performance excellence.

EXERCISE

Type of Activity: Large Group

Instructions

1. Evaluate each statement below to determine whether the performance is behaviorally stated and measurable.
   a. Marian was dressed sloppily.
   b. John omitted the 1900 hour dose of Metaclopramide.
   c. Beth felt that the family was displeased with their mother's care.
   d. Jerry was late for shift report three times this week.
   e. Laura's charting was inadequate.
   f. Bob is an excellent employee.
   g. The narcotic count was incorrect five times this month.
   h. Jenny has attended six continuing education offerings.
   i. Dan appears competent when delivering care.
   j. Emma has great rapport with the patients.
2. As a class, propose a scenario where a positive event occurs, and then a negative event occurs.
3. Describe the event using behaviorally-based language that is measurable.

Discussion

1. From this exercise, what did you learn about writing in behavioral terms?

Further Reading

Jasper, M. A. (1995). The potential of the professional portfolio for nursing maintenance and verification of continuing nursing practice following initial nurse training in the UK. *Journal of Clinical Nursing*, 4(4), 249–255.

Rosenthal, L. (1995). Exploring the learner's world: Critical incident methodology. *Journal of Continuing Education in Nursing*, 26(3), 115–118.

MODULE 18 STAFF DEVELOPMENT

Exercise 18–1   Coaching as a Staff Development Strategy

Exercise 18–2   Providing a Formal Performance Appraisal

Exercise 18–3   Multicultural Issues in Staff Development

■ Exercise 18–1 Coaching as a Staff Development Strategy

INTRODUCTION

An important component of the continuous quality improvement strategy is to institute on-the-job training. Indeed, the rapid pace of change in health care requires that employees, including management, must be constantly educated on new technologies, policies, and procedures. Therefore, a manager must be able to efficiently transmit this knowledge in an easily understood and accepted manner.

One way to provide this knowledge is to provide coaching. Coaching is the day-to-day, proactive process of helping employees improve their performance. It is often a neglected activity of nurse managers that results in frustrated employees. Coaching is also used when the employee possesses the needed knowledge, but can be assisted in achieving higher skill performance. Prior to the coaching session, the manager should prepare for the meeting by assessing the employee's education, existing skills, ability to perform the activity, and his or her readiness to learn. The steps of the coaching process are listed on page 143.

1. State the targeted performance in behavioral terms.
2. Tie the problem to consequences and to a pertinent outcome, so that the employee knows why the behavior is a problem. Avoid use of passive or aggressive language—state your assessment assertively in behavioral terms.
3. Having stated the problem behavior, listen openly as the employee describes the problem and the reasons for it.
4. Seek input from the employee on how to solve the problem.
5. Mutually decide on what behavioral actions will be taken to solve the problem. Document the solution for later reference.
6. Schedule a follow-up meeting when the employee can receive performance feedback.

EXERCISE

TYPE OF ACTIVITY: SMALL GROUP

Instructions

1. Read the scenarios.
2. In groups of three, role-play the following scenarios. Take turns being the coach, and have the other two group members play the employee being coached and an observer.

Scenarios

1. You are the manager of a cardiac intensive care unit. You are planning to coach the telemetry technician who has been struggling with the documentation of the cardiac telemetry strips.
2. You are the case manager for home health care clients. You have identified a problem with one of the home health aides working with Mr. Gavin. Mr. Gavin states that the aide did not provide catheter care last evening. You plan to meet with the aide to discuss the problem.
3. You are a senior nurse at the community renal dialysis center. You have been orienting a new RN to the unit and notice that this RN is uncomfortable with her intravenous catheter insertion technique. Since this is a vital skill for this job, you schedule an opportunity to give her some pointers.

Discussion

1. What was the most difficult part of the coaching process?
2. What is the difference between coaching and teaching? Between coaching and discipline?
3. What are the issues involved in diagnosing a performance problem?

Further Reading

Haas, S. A. (1992). Coaching. Developing key players. *Journal of Nursing Administration, 22*(6), 54–58.

Wise, P. S. Y. (1992). Learning needs assessment. In R.S. Abruzzese (Ed.), *Nursing Staff Development: Strategies for Success* (pp. 183–214). St. Louis, MO: Mosby Year Book.

Exercise 18-2 Providing a Formal Performance Appraisal

INTRODUCTION

An important component of staff development is the formal performance appraisal. Appraisals are given at least yearly and provide ongoing constructive feedback of employee performance. The appraisal is future-oriented and structured to enhance the productivity of the employee.

Appraisal accuracy is a primary concern for the manager. Questionable performance appraisals have resulted in employee-employer lawsuits. To ensure that the appraisal system is non-discriminatory, the steps listed below should be present in the organization's appraisal system.

1. The appraisal should be in writing and carried out at least once a year.
2. The performance appraisal information should be shared with the employee.
3. The employee should have the opportunity to respond in writing to the appraisal.
4. Employees should have a mechanism to appeal the results of the performance appraisal.
5. The manager should have adequate opportunity to observe the employee's job performance during the course of the evaluation period. If adequate contact is lacking, then appraisal information should be gathered from other sources.
6. Anecdotal notes ("critical incidents") on the employee's performance should be kept during the entire evaluation period and shared at the time of the employee's appraisal.
7. Evaluators should be trained to carry out the performance appraisal process.
8. The performance appraisal should focus on employee behavior and results rather than on personal traits or characteristics (such as attitude or personality).

The performance appraisal should be provided in a positive, non-threatening manner. As you participate in the following scenario, practice the key behaviors listed below.

1. Put the person at ease.
2. Make it clear that the purpose of the performance appraisal is to help the employee achieve maximum performance.
3. Review the outcome ratings with the employee, citing specific examples of behavior that resulted in a particular rating.
4. Mutually determine future performance goals and document them on the performance appraisal form.
5. Set a follow-up date within three months.
6. Express your confidence in the employee.

EXERCISE

TYPE OF ACTIVITY: SMALL GROUP

Instructions

1. Divide into groups of three. In each group, one member is designated to play the part of Ryan, one member is designated as the nurse manager, and one member is the observer.
2. Participants from each group who play Ryan leave the room. While waiting, they review the role-play for Ryan on page 147 and the Case Study on page 148.
3. Participants from each group who play the nurse manager remain in the room and review the Case Study, the Performance Appraisal Form (Figure 1 on page 149), and legal criteria for performance appraisals (page 145).
4. The nurse managers complete the Performance Appraisal.
5. Group members who are role-playing Ryan return to their respective groups and the performance appraisal is conducted.
6. Observers provide feedback based on the behaviors outlined above.

Discussion

1. What steps in the performance review were most difficult?
2. As the nurse manager, what other information would be helpful to improve the performance appraisal?

Further Reading

Murphy, K. R. & Cleveland, J. N. (1991). *Performance Appraisal.* Boston: Allyn & Bacon.

McGee, K. G. (1992). Making performance appraisals a positive experience. *Nursing Management,* 23(8), 36–37.

Tayler, C. M. (1992, September/October). Subordinate performance appraisal: What nurses really want in their managers. *Canadian Journal of Nursing Administration,* pp. 6–9.

ROLE-PLAY FOR RYAN, RN

You had worked as a RN in acute care settings for five years prior to changing jobs. This is your first job outside of a hospital. As a nurse educator in a public health department, you are learning new skills every day. Your main interest is in providing quality patient education and you feel that you are capable of meeting this goal. You have always succeeded in achieving your goals, and believe that you will be successful in this new field. You are anxious, however, about this performance appraisal—your supervisor has only occasionally observed your performance and you do not believe that she has the necessary information to provide a knowledgeable appraisal.

Your expectations of yourself and others tends to be very high. You are highly motivated to excel and have little tolerance for those who are not as motivated. You feel, in fact, that most nurses do not measure up to your expectations of the professional nurse. To be perfectly honest, you have been disappointed by the lack of challenge in this position, and you seek to move into management as soon as possible. You sense a certain tension between yourself and your manager. You are distrustful of most managers and suspect that your manager will rate you incorrectly. You challenge her ratings openly and try to avoid the supervisor's attempts to set goals in the areas that need improvement.

NURSE MANAGER CASE STUDY, RYAN, RN

Ryan has been employed by the public health department for one year. This is his first performance appraisal. Ryan shows promise as a nurse educator. He is knowledgeable and has shown initiative in creating new educational materials for the department.

Ryan has been very professional in his demeanor and dress. He is always punctual, has had near-perfect attendance and interacts well with the physicians. When assigned new activities, Ryan carries them out correctly. He takes the initiative and looks for opportunities to improve the department. Sometimes he goes too far, however, in spending too much time with each patient. This has decreased his productivity in terms of patients seen per hour. When you mentioned this to Ryan, he appeared defensive. In general, you feel that Ryan has difficulty taking criticism.

Since your first discussion with Ryan about this issue, the other nurse educator has complained to you about Ryan's productivity. She has had to pick up the slack, seeing more than her share of patients because of Ryan's increased time with his patients. This has caused some friction between them.

Ryan's attention to detail shows in his precise documentation. On one occasion, however, you witnessed him criticizing the other educator's documentation in front of the secretarial staff. Because of this and other similar situations where staff has been critiqued by Ryan, the other nurses and staff are beginning to avoid him.

FIGURE 1 Health Department Performance Appraisal Form

Complete the following form. In the section for comments, write specific reasons for your ratings.

	Achieves Stated Outcome			
	25%	50%	75%	100%
1. Quality. Meets expectations for thoroughness and accuracy. Makes sound decisions based on facts. Works within hospital policies and procedures.				X
2. Quantity. Makes efficient use of work time. Meets standards for rate of progress on her assigned tasks.				X
3. Knowledge. Understands all parts of the position and how it relates to other positions. Keeps informed of current events in the field.				X
4. Ability to work with others. Maintains respectful and effective relationships with all levels of personnel. Has a cooperative attitude toward patients and visitors.			X	
5. Attendance. Works scheduled days.				X
6. Punctuality. Reports on duty as expected.				X
7. Initiative. Requires little supervision. Solves problems and takes responsibility when needed.				X

Recommend employee for merit pay increase   Yes _____   No _____

Comments on Specific Situations:

4. Follow appropriate channels of communication 1st supervisor 2nd manager - let them correct ~~their~~ your peers

The ↓ in # of pts. is not worthy of criticism

Exercise 18-3 Multicultural Issues in Staff Development

INTRODUCTION

Nurse managers must consider cultural differences when planning staff development. The astute manager includes a cultural assessment in his or her needs assessment. Cultural issues to consider include socialization differences, cultural work expectations, and language equivalence. In planning staff development, the manager should investigate any cultural specifics that will enhance the employee's receptivity and understanding of the development program. Three issues need to be addressed:

1. *Functional equivalence*   Is the problem identified by the manager perceived to be a problem by the ethnic group of the employee?

2. *Comparative descriptive framework*   Do the developmental strategies provide illustrations that allow for cross-cultural conceptualizations of the behavior being modeled?

3. *Language equivalence*
    - Do conceptual definitions have the same meaning in both cultures?
    - Do operational indicators appropriately reflect the conceptualizations for individuals in another culture?
    - Given bilingual, bicultural individuals, do items written in two languages have the same meaning?

EXERCISE

Type of Activity: Large Group

Instructions

Read and discuss the following scenario.

Scenario

An elderly patient is dying at home per request of the patient and the family. As the home health nurse managing the case, you need to educate the home health aide about the process of preparation of a body in the event that the aide is present when the patient expires. You wish to be culturally sensitive.

Discussion

1. Discuss cultural perspectives about the death experience.
2. What are the issues that the nurse case manager must consider as he or she provides information to the home health aide?

Further Reading

Burner, O. Y., Cunningham, P. & Hattar, H. S. (1990). Managing a multicultural nurse staff in a multicultural environment. *Journal of Nursing Administration,* 20(6), 30-34.

Phillips, L. R., de Hernandez, I. L. & de Ardon, E. T. (1994). Strategies for achieving cultural equivalence. *Research in Nursing and Health,* 17, 149–154.

MODULE 19 — MOTIVATION

Exercise 19–1 Personal Motivation: Use of Goal Setting
Exercise 19–2 Employee Motivation

■ Exercise 19–1 Personal Motivation: Use of Goal Setting

INTRODUCTION

Personal motivation is the drive to perform at your best. One way to motivate yourself is to set goals. Goal-setting theory has three basic propositions: (1) specific goals lead to higher performance than do general goals; (2) specific, difficult goals lead to higher performance than specific, easy goals provided the goals are accepted by those involved; and (3) incentives such as money, knowledge of results, praise and reproof, participation, competition, and time limits affect behavior only if they cause individuals to change their goals or to accept goals that have been assigned to them. As a nurse manager, you have the opportunity to assist staff in setting goals to enhance their professional performance.

EXERCISE

TYPE OF ACTIVITY: SMALL GROUP

Instructions

1. Divide into pairs and read the scenario.
2. Discuss the goals that might be appropriate for this individual.
3. Complete the attached goal setting tool (Figure 1).

FIGURE 1 Goal Setting

	1 year	5 years	10 years
Professional goals			
Educational goals			
Work focus			
Professional activities			

Scenario

Assume that you are the nurse manager of a home health care agency. You have just hired a student nurse to work as a home health aide. The student is very anxious about the job responsibilities of working alone in a patient's home. You recognize that the student might benefit by setting some goals for professional growth in this position.

Discussion

1. Were there areas where goals were easy to set? Difficult to set?
2. How will you assist the home health aide in achieving these goals? Be as specific as possible.

Further Reading

Griffiths, P. (1995). Progress in measuring nursing outcomes. *Journal of Advanced Nursing*, 21(6), 1092–1100.

Exercise 19-2 Employee Motivation

INTRODUCTION

Motivation theory focuses on methods to direct employee behavior. Motivation literally means "to move." Management has always been interested in motivating employees to achieve their potential, leading to increased productivity. You must recognize, however, that as unique individuals, employees are motivated by different things. It is this fact that makes motivation a challenging and interesting management skill.

Many theories have been developed to understand motivation. Some of the better known are Maslow's theory of human motivation, Herzberg's two factor theory which relates motivation to job satisfaction, Skinner's reinforcement theory, Vroom's expectancy theory, and Locke's goal-setting theory. Theory application is useful when learning how to motivate. You are encouraged to refer to the text for further information on these theories.

It is important to understand *what* motivates individuals. Motivation can be intrinsic or extrinsic. *Intrinsic motivation* comes from within the individual; *extrinsic motivation* is derived from the work setting. Both types of motivation are associated with one's values and belief system. Factors such as money, social status, group harmony, or praise from a superior may extrinsically motivate an individual. Factors such as sense of accomplishment of difficult tasks or improvement of self-knowledge may intrinsically motivate the employee. It is the responsibility of an astute manager to identify the motivating factors of an individual and to help that individual attain their work potential using these factors.

EXERCISE

Type of Activity: Small Group

Instructions

1. Interview three different individuals working in health care who perform different types of work at different levels of the organizational hierarchy. On a scale of 1–8 with 8 being the most motivating, ask them to prioritize the factors listed on the motivation priority list (Table 1).
2. Compile their responses on Table 1 and bring the completed tool to class.
3. In class, divide into groups of three and share your results.
4. Answer the discussion questions.

TABLE 1	MOTIVATION PRIORITY LIST		
MOTIVATIONAL FACTOR	INDIVIDUAL 1	INDIVIDUAL 2	INDIVIDUAL 3
■ Job title			
■ Knowledge			
■ Financial gain			
■ Social status			
■ Group harmony			
■ Praise			
■ Career advancement			
■ Sense of accomplishment of difficult tasks			
■ Fear of discipline			
■ Other			

Discussion

1. Do you see any similarities or differences in motivators between levels of the organization?

2. Do you see any similarities or differences in motivators between different professions?

3. Based on what you have discovered, what suggestions would you have about extrinsically motivating individuals?

4. How can you accomplish this?

Further Reading

Dealy, M. F. & Bass, M. (1995). Professional development: Factors that motivate staff. *Nursing Management,* 26(8), 32F–32I.

Henderson, M. C. (1993). Measuring managerial motivation: The management inventory. *Journal of Nursing Measurement,* 1(1), 67–80.

SECTION VII
IDENTIFYING ETHICAL AND LEGAL ISSUES

MODULE 20   **AVOIDING MALPRACTICE**

Exercise 20–1   Identifying Malpractice
Exercise 20–2   Comparing Standards
Exercise 20–3   Developing Policies and Procedures

MODULE 21   **MINIMIZING ORGANIZATIONAL RISK**

Exercise 21–1   Handling Patient and Family Complaints
Exercise 21–2   Understanding Advanced Directives
Exercise 21–3   Understanding the Role of the State Board of Nursing

MODULE 22   **LEGALITIES IN WORKING WITH PERSONNEL**

Exercise 22–1   Reducing Liability in Hiring Decisions
Exercise 22–2   Understanding Union Grievance Hearings
Exercise 22–3   Developing a Grievance Policy and Procedure

MODULE 20 AVOIDING MALPRACTICE

Exercise 20–1	Identifying Malpractice
Exercise 20–2	Comparing Standards
Exercise 20–3	Developing Policies and Procedures

■ Exercise 20–1 Identifying Malpractice

INTRODUCTION

Nurses are legally liable for their actions. Failure to function as a prudent nurse is *negligence,* which is referred to as malpractice in the case of professionals. In order for malpractice to exist four elements must be present: duty, breach of duty, causation, and injury. *Duty* is the responsibility of a professional to act in a particular manner. *Breach of duty* occurs when an individual either performs an act (commmission) leading to harm, or fails to perform (omission) as a reasonably prudent professional would. The breach of duty *causes* an *injury* to occur.

EXERCISE

TYPE OF ACTIVITY: INDIVIDUAL OR SMALL GROUP

Instructions

1. Analyze the following scenario for evidence of
    a. the nurse's duty
    b. a breach of duty
    c. an injury
    d. causation
2. Identify whether evidence for malpractice exists.

Scenario

Barbara is a RN with 25 years of experience. She works in a small community hospital where she is frequently the only nurse on nights, which means she covers the emergency room and any other area of the hospital as needed. The hospital primarily cares for elderly patients. Pediatric and obstetric patients are generally referred to an urban hospital about 20 miles away. Emergencies are triaged, stabilized, and transported as soon as possible. Recently a woman who had had no prenatal care arrived at the emergency room too late in labor to transport. Fortunately, the physician on call arrived just in time to deliver the infant. The infant was apneic at birth with a weak heart rate of 64. After some searching, Barbara found an infant resuscitation bag and started ventilating the infant using adult rates, since she wasn't aware of any differences for infants and room air (21% FiO_2). The infant steadily worsened and died.

Analysis/Discussion

1. What is a nurse's responsibility for maintaining competence in different specialties? How do you feel about this?

2. Do you think Barbara is guilty of malpractice? Why or why not?

Further Reading

Aiken, T. D. (1994). *Legal, Ethical, and Political Issues in Nursing*. Philadelphia, PA: F. A. Davis.

Fiesta, J. (1994). Nursing torts: From plaintiff to defendant. *Nursing Management*, 25(2), 17–18.

Exercise 20-2 Comparing Standards

INTRODUCTION

A standard is a criterion for measuring performance. Explicit, measurable terms are used to establish a level of excellence. Standards guide practice, define quality, and reduce risk. There are three types of standards: outcome, process, and structure. *Structure* standards establish criteria relevant to the organization such as staffing criteria, cost constraints, and environmental safety. *Process* standards are task-oriented and define how nursing care should be delivered. *Outcome* standards or measures are used to determine if the desired results have been achieved. In nursing, outcome measures reflect patient objectives.

EXERCISE

Type of Activity: Individual

Instructions

1. Obtain a copy of each of the following:
   a. Joint Commission on Accreditation of Healthcare Organizations (JCAHO) standards.
   b. American Nurses Association standards.
   c. A specialty nursing organization's standards.
   d. A copy of an organization's policy that corresponds to the selected specialty organization.
2. Answer the analysis questions.

Analysis

1. Is the organization's policy based on a JCAHO, ANA, or specialty organization standard?
2. How are JCAHO and ANA standards similar? How are they different?
3. How similar and dissimilar are the specialty nursing organization's standards with those of the ANA?

Further Reading

Luquire, R. et al. (1994). Focusing on outcomes. *RN*, 57(5), 57–60.
Mitchell, M. K. (1989). The power of standards: The glory days of nursing yet to come? *Nursing and Health Care*, 10(6), 306–309.

Exercise 20-3 Developing Policies and Procedures

INTRODUCTION

Policies and procedures are guides for accomplishing goals and objectives. *Policies* are comprehensive, general statements used to guide decision making and to maintain consistency. *Procedures* provide specific directions or steps for implementing a policy and are frequently a tool for standardization and evaluation.

EXERCISE

TYPE OF ACTIVITY: INDIVIDUAL

Instructions

1. Write a policy about absenteeism.
2. Write a procedure for reporting absences and the consequences of repeated absences.

Analysis

1. What types of considerations were necessary in writing this policy?
2. Were there aspects of absenteeism or attendance you had not considered before? What were they?
3. Evaluate your procedure for fairness and effectiveness. What are the legal implications?

MODULE 21 — MINIMIZING ORGANIZATIONAL RISK

Exercise 21–1   Handling Patient and Family Complaints

Exercise 21–2   Understanding Advanced Directives

Exercise 21–3   Understanding the Role of the State Board of Nursing

■ Exercise 21–1 Handling Patient and Family Complaints

INTRODUCTION

Handling patient and family complaints is an important step in reducing the risk of lawsuits. It is often the dissatisfied patient or family that pursues litigation. When a complaint is voiced it is important to use active listening techniques and to avoid becoming defensive or emotional. Let the person describe their concerns before speaking. Elicit the individual's expectations and clarify what you can or cannot do. Maintain a caring attitude as you negotiate actions to be taken and decide upon a time frame.

EXERCISE

TYPE OF ACTIVITY: INDIVIDUAL OR SMALL GROUP

Instructions

1. For each of the following scenarios,
    a. Identify the risk factor(s).
    b. Describe how you would handle the situation.

Scenarios

1. Estella Rice is a 96-year-old resident in your long-term care facility. Mrs. Brown, her daughter, comes to you complaining that they cannot find her mother's brand new pink nightgown and robe set anywhere.

2. Mrs. Rodriquez comes to you, the clinic manager, to complain about a male pediatric resident who examined her 12-year-old daughter. She felt he had been inappropriate by "touching her chest with his hand."

3. Mr. Thomas, a hospital visitor, says he just went out to his car and discovered his attache case was missing.

4. Mr. Chueng fell on the ice in the hospital parking lot and fractured his elbow.

Analysis/Discussion

1. What was the easiest scenario to identify the risk involved? The most difficult?

2. What was your initial response to each of the scenarios?

3. How easy do you feel it was to avert risk in each situation?

Further Reading

Goldman, T. A. (1991). Risk management concepts and strategies. *Journal of Intravenous Nursing,* 14, 199–204.

Exercise 21-2 Understanding Advanced Directives

INTRODUCTION

The Patient Self-Determination Act passed in 1991 requires all institutions receiving Medicare or Medicaid funds to provide patients with written information about their rights under state law to make decisions about their medical care. These rights include the right to initiate advanced directives such as living wills and durable power of attorney (Figure 1). *Advanced directives* provide a mechanism for communicating health care treatment preferences when the capacity for decision making is lost. *Living wills* are used to communicate wishes regarding life-sustaining procedures, while the *durable power of attorney* appoints an individual to make decisions regarding treatment that may not be covered in the living will.

EXERCISE

TYPE OF ACTIVITY: INDIVIDUAL OR SMALL GROUP

Instructions

1. See the example of an advanced directive and a durable power of attorney in Figure 1.

2. Discuss the role of each for the patient giving an advanced directive.

3. Evaluate the language of the advanced directive. Are there any gray areas? What types of ethical dilemmas will those gray areas present for health care providers and family members?

4. Read the following scenario and discuss the ethical-legal dilemmas present.

Scenario

Ali Mohamed had multiple sclerosis. While in the nursing home, he fell and broke his elbow. He was taken to the hospital for treatment. He was admitted for overnight rehydration with IV fluids and a decrease in swelling before surgery was attempted. Mr. Mohamed is 56 years old and has filled out an advanced directive, but not a durable power of attorney. In the advanced directive, it is stipulated that Mr. Mohamed is not to be resuscitated nor kept alive through artificial means.

That night Mr. Mohamed became confused and disoriented. He pulled out his IV. The next day he refused to eat. Because of his mental status, surgery was postponed. Mr. Mohamed continued to be confused and continued to refuse to eat. In the meantime, his mental status

FIGURE 1 Sample Durable Power of Attorney for Health Care Decisions

Health Care Treatment Directive

I _____ make this Health Care Treatment Directive to exercise my
        Print Name
my right to determine the course of my health care and to provide clear and convincing proof of my treatment decisions when I lack the capacity to make or communicate my decisions and there is no realistic hope that I will regain such capacity.

If my physician believes that a certain life prolonging procedure or other health care treatment may provide me with comfort, relieve pain or lead to a significant recovery, I direct my physician to try the treatment for a reasonable period of time. However, if such treatment proves to be ineffective, I direct treatment be withdrawn even if so doing may shorten my life.

I direct I be given health care treatment to relieve pain or to provide comfort even if such treatment might shorten my life, suppress my appetite or my breathing, or be habit-forming.

I direct all life prolonging procedures be withheld or withdrawn when there is no hope of significant recovery, and I have:
- a terminal condition; or
- a condition, disease or injury without reasonable expectation that I will regain an acceptable quality of life; or
- substantial brain damage or brain disease which cannot be significantly reversed.

1. When any of the above conditions exist, I DO NOT WANT the life prolonging procedures which I have initialed below. (You should assume any treatments not initialed may be administered to you.)
   - surgery .... _____ initials
   - heart-lung resuscitation (CPR) .... _____ initials
   - antibiotics .... _____ initials
   - dialysis .... _____ initials
   - mechanical ventilator (respirator) .... _____ initials
   - tube feedings (food and water delivered through a tube in the vein, nose, or stomach) .... _____ initials
   - other _____ _____ initials

2. I make other instructions as follows: (You may describe what a minimally acceptable quality of life is for you.)

   _____
   _____
   _____
   _____

If you do not wish to name an agent as referred to on the reverse side, initial here _____ , write "None" in the space provided for agent's name, sign and have witnessed and/or notarized.

FIGURE 1 *continued*

This is a Durable Power of Attorney for Health Care Decisions, and the authority of my agent shall not terminate if I become incapacitated. I grant my agent full authority to make decisions for me regarding my health care. In exercising this authority, my agent shall follow my desires as stated in my Health Care Treatment Directive or otherwise known to my agent. My agent's authority to interpret my desires is intended to be as broad as possible and any expenses incurred should be paid by my resources. My agent may not delegate the authority to make decisions. My agent is authorized as follows to:

> If there is a statement in paragraphs 1 through 6 below with which you do not agree, draw a line through it and add your initials.

1. Consent, refuse or withdraw consent to any care, treatment, service or procedure, (including artificiality supplied nutrition and/or hydration/tube feeding) used to maintain, diagnose or treat a physical or mental condition;
2. Make decisions regarding organ donation, autopsy and the disposition of my body;
3. Make all necessary arrangements for any hospital, psychiatric hospital or psychiatric treatment facility, hospice, nursing home or similar institution; to employ or discharge health care personnel (any person who is licensed, certified or otherwise authorized or permitted by the laws of the state to administer health care) as the agent shall deem necessary for my physical, mental and emotional well being;
4. Request, receive and review any information, verbal or written, regarding my personal affairs or physical or mental health including medical and hospital records and to execute any releases of other documents that may be required in order to obtain such information;
5. Move me into or out of any state for the purpose of complying with my Health Care Treatment Directive or the decisions of my agent;
6. Take any legal action reasonably necessary to do what I have directed.

I appoint the following person to be my agent to make health care decisions for me WHEN AND ONLY WHEN I lack the capacity to make or communicate a choice regarding a particular health care decision and my Health Care Treatment Directive does not adequately cover circumstances. I request that the person serving as my agent be my guardian if one is needed.

> If you do not wish to name an agent, write "None" in the space provided below.

Agent's Name: _____ Telephone: _____

Address: _____

If my agent is not available or not willing to make health care decisions for me or, if my agent is my spouse and is legally separated or divorced from me, I appoint the person or persons named below (in the order named if more than one listed) as my agent: (It is not necessary to name an alternate agent.)

First Alternate Agent   Second Alternate Agent

Name: _____   Name: _____

Address: _____   Address: _____

Telephone: _____   Telephone: _____

Protection of Persons Who Rely on My Agent: I and my estate hold my agent and my caregivers harmless and protect them against any claim for following this durable power of attorney.

Severability: If any part of this document is held to be unenforceable under law, I direct that all of the other provisions of the document shall remain in force and effect.

Date: _____ X Signature _____

Witness _____ Date _____ Witness _____ Date _____

Note: Used with the permission of the Midwest Bioethics Center, Kansas City, MO.

deteriorated. Since Mr. Mohamed was no longer able to make decisions regarding his health care, the medical staff approached Mrs. Mohamed about inserting a feeding tube. Mrs. Mohamed was hesitant since Mr. Mohamed had indicated he did not want to be kept alive through artificial means. The physicians on the case did not view nasogastric tube feedings as a means to keep Mr. Mohamed alive artificially. However, they agreed to talk with other family members to better appreciate Mr. Mohamed's wishes. The only other family member was a nephew who was a paramedic. He reinforced the fact that Mr. Mohamed did not want to be kept alive artificially, but agreed to having a feeding tube inserted for a three-day period. Before feedings could be started, Mr. Mohamed had a respiratory arrest and died.

Analysis/Discussion

1. What role did the advanced directive play in guiding the care of Mr. Mohamed?
2. Did the medical staff follow the advanced directive?
3. Did the family have a right to make medical decisions about Mr. Mohamed's care?
4. Is there evidence of malpractice in this case?
5. How could ethical-legal problems in this case be prevented?

Further Reading

American Nurses Association (1991). *Nursing and the Patient Self-Determination Act.* Washington, DC: ANA.

Exercise 21-3 Understanding the Role of the State Board of Nursing

INTRODUCTION

Each state has a board of nursing that ensures that its nurse practice acts are carried out. Nurse practice acts define the scope of nursing, how licensure is obtained or revoked, and the penalties for practicing without a license, as well as the make-up of the nursing board and its responsibilities. The board of nursing is usually responsible for reviewing the nurse practice act to ensure minimum standards of practice are defined, approving nursing education programs, evaluating applicants and issuing nursing licenses, and disciplining nurses who violate the law. Most disciplinary action is related to substance abuse. For this reason, many states provide diversionary programs that assist nurses in overcoming their substance abuse and returning to practice.

EXERCISE

TYPE OF ACTIVITY: SMALL OR LARGE GROUP

Instructions

1. Attend a state board of nursing meeting.
2. Be prepared to discuss the discussion questions.

Discussion

1. What are the professional backgrounds of those on the board of nursing?
2. What types of issues does the board address?
3. Does the board offer a diversion or peer program for substance abusers?
4. Under what conditions is a nursing license suspended? revoked?

Further Reading

American Nurses Association (1990). *Suggested State Legislation Nursing Practice Act, Nursing Disciplinary Act, Prescriptive Authority Act.* NP78, Kansas City, MO: The American Nurses Association.

MODULE 22 — LEGALITIES IN WORKING WITH PERSONNEL

Exercise 22—1	Reducing Liability in Hiring Decisions
Exercise 22–2	Understanding Union Grievance Hearings
Exercise 22–3	Developing a Grievance Policy and Procedure

■ Exercise 22–1 Reducing Liability in Hiring Decisions

INTRODUCTION

A number of laws govern hiring practices. One of the first of these was the *Civil Rights Act of 1964 (Title VII)*. Title VII prohibits discrimination on the basis of race, color, sex, or national origin. Two exceptions are bona fide occupational qualifications and bona fide seniority or merit systems. If bona fide occupational qualifications exist, employment decisions may be made on the basis of national origin, religion, or sex. Likewise, it is permissible to make decisions about promotions and lay-offs based on seniority. In 1967 the *age discrimination in employment act* (ADEA) was passed that made it unlawful for employers to discriminate against older men or women. This law was extended in 1986 to prohibit discrimination of individuals over the age of 40 and to prohibit mandatory retirement for persons under the age of 70. In 1990 the *Americans with Disability Act* (ADA) was passed, prohibiting discrimination against disabled persons.

EXERCISE

TYPE OF ACTIVITY: INDIVIDUAL

Instructions

1. Read the following scenario.
2. Address the questions in the analysis section.

Scenario

You have placed an advertisement in the local newspaper for a staff nurse for your home health care agency. The following individuals apply:

Gary is a 50-year-old European American male who recently completed a LPN program. He has no prior health care experience. In fact, Gary has minimal short-term work experience and many gaps in his employment history.

Marcia is a 28-year-old African-American female with seven years of experience as a RN in pediatrics. She openly informs you she is a single parent of two daughters and is interested in home care for its hours.

Cynthia is a 22-year-old, European American female with one year of experience as a RN in a physician's office.

Juarez is a 40-year-old Latino with 20 years of experience as a Registered Nurse in Oncology and Hospice. He has an excellent work history. When asked why he is considering a job change, he responds, "I'm tired of the gay-bashing at work."

Analysis

1. Who would you hire? Why?
2. What issues may be raised by those you don't choose?

Further Reading

Wray, R. S. and O'Connor, N. M. (1990). Employers and the disability act. *Business Health,* 8(10), 66.

Exercise 22-2 Understanding Union Grievance Hearings

INTRODUCTION

Grievance proceedings serve to protect an individual's or work group's rights. Grievances are usually the result of personal problems, union politics, or unfavorable contract language. Grievance procedures provide a forum for reviewing and resolving a dispute within a specific time frame. The first step in any grievance is to informally discuss the situation with the immediate supervisor. If the grievance cannot be satisfactorily resolved, formal negotiations are initiated in writing by the employee. A written response is provided by administration. If a satisfactory response is not obtained, the grievant and his or her agent meet with the organization's grievance chairperson and an administrator. A written response is provided following the meeting. If a satisfactory agreement is not reached, the next step is to include the director of human resources in the discussions. If the answer is still not satisfactory, then the final step is arbitration. In arbitration both sides select a representative. The representatives review the grievance and research the situation. These representatives present their findings to an arbitrator who studies the evidence and makes a final decision.

EXERCISE

TYPE OF ACTIVITY: SMALL OR LARGE GROUP

Instructions

1. The instructor will seek volunteers to role-play Armand, the grievant; a bargaining unit representative; an arbitrator; a manager; and a director of human resources. The rest of the class are observers.
2. Conduct a grievance hearing.
3. The observers should note
    a. the type of communication that is most productive.
    b. the behavior that is most conducive to negotiation.

Scenario

Armand is upset at being passed over for a merit raise. He talks with his manager, but is not satisfied with the response and initiates grievance proceedings.

Discussion

1. What types of behavior facilitated negotiation of Armand's grievance?
2. How is negotiation in union proceedings different from or similar to negotiation in individual conflict?

Further Reading

Wilson, C. N., Hamilton, C. L. & Murphy, E. (1990). Union dynamics in nursing. *Journal of Nursing Administration,* 20, 35–39.

Exercise 22-3 Developing a Grievance Policy and Procedure

INTRODUCTION

A mechanism for handling grievances is also important in settings without collective bargaining units. Having a grievance policy and procedure can help reduce organizational risks and enhance employee satisfaction.

EXERCISE

TYPE OF ACTIVITY: INDIVIDUAL

Instructions

1. Read the following scenario.
2. Write a grievance policy and procedure for County Hospital.

Scenario

County Hospital recently had a change in management. The new administrator was appalled to learn the hospital has never had a grievance policy or procedure for its employees. The administrator has asked all managers and any interested employees to submit such a policy. A policy and procedure will be developed from the suggestions given.

Analysis

1. How easy/difficult was it to write the grievance procedure?
2. What did you learn about grievance procedures?

SECTION VIII: SURVIVING IN A CHANGING HEALTH CARE ENVIRONMENT

MODULE 23   **PERSONAL IMAGE AND CAREER DEVELOPMENT**

Exercise 23–1   Developing a Résumé
Exercise 23–2   Developing a Career Plan

MODULE 24   **MANAGING CHANGE**

Exercise 24–1   Developing Strategies for Change
Exercise 24–2   Managing Resistance

MODULE 23 — PERSONAL IMAGE AND CAREER DEVELOPMENT

Exercise 23–1   Developing a Résumé

Exercise 23–2   Developing a Career Plan

Exercise 23–1   Developing a Résumé

INTRODUCTION

A résumé is a concise and accurate history of education and work-related experiences. Since résumés are often your first impression on a prospective employer, its appearance should be neat, attractive, easy-to-read, and professional. Above all, the résumé should maximize your strengths and market you.

EXERCISE

Type of Activity: Individual

Instructions

1. Identify a position in a newspaper, journal, or other job source.
2. Write a cover letter expressing your interest in the job.
3. Develop a résumé.

Analysis

1. What areas were difficult to write? Why?
2. Trade résumés with a friend. How does your résumé compare? Which is more aesthetically pleasing? Who has a better experiential background? What first drew your attention in the other résumé?

Further Reading

Dadich, K. A. (1992). Your résumé. *Healthcare Trends and Transitions*, 3(2), 20–21, 96.

Exercise 23-2 Developing a Career Plan

INTRODUCTION

Success and satisfaction in a career are attained by careful career planning. Career planning is an ongoing deliberate process involving (a) self-assessment of interests, skills, abilities, and strengths and weaknesses; (b) identification of goals and priorities; and (c) determination of strategies.

EXERCISE

TYPE OF ACTIVITY: INDIVIDUAL

Instructions

1. Answer the following questions.
    a. What area of nursing would you like to work in?
    b. Would you like to go on for an advanced degree?
    c. Are you interested in an expanded nursing role?
    d. How do family and social plans interact with your career goals?
    e. What are your strengths? your weaknesses?
    f. Are you interested in writing? research? giving presentations?
2. Identify the importance of family, profession, and social life by assigning a percentage to each life priority. Together, these life priorities should total 100 percent. Develop a career map that considers the answers to these questions and the priorities you have identified using the Career Map Guide in Table 1. The map should include 1-, 5-, and 10-year objectives for career, education, professionalism, and personal endeavors.

    Family   _____%
    Profession   _____%
    Social life   _____%

TABLE 1	CAREER MAP GUIDE		
Objectives	1 year	5 years	10 years
■ Professional experiences			
■ Education (formal and informal)			
■ Community involvement			
■ Scholarly activities (research, publications, etc.)			

Analysis

1. Did you discover anything new about your priorities in this exercise?
2. How difficult or easy was it to develop objectives?

Further Reading

Vogel, G. (1990). Career development: An integrated process. *Holistic Nursing Practice,* 4(4), 46–53.

MODULE 24 MANAGING CHANGE

Exercise 24–1   Developing Strategies for Change

Exercise 24–2   Managing Resistance

Exercise 24–1   Developing Strategies for Change

INTRODUCTION

Change is a dynamic process in which an imbalance between driving and restraining forces moves behavior to a new level where the forces reach a new state of equilibrium. Common strategies used to evoke change are power-coercive, empirical-rational, and normative-reeducative. In *power-coercive* strategies legitimate authority, economic sanctions, or political clout are used to enforce change and handle resistance. The premise behind the *empirical-rational* approach is that people are rational and will accept a change that is in their best interest. The *normative-reeducative* approach is based on the assumption that people respond to social norms and values. Hence, roles, relationships, perceptual orientations, attitudes, and feelings influence the acceptance of change. Agents using a normative-reeducative approach collaborate with the group to effect change.

EXERCISE

TYPE OF ACTIVITY: INDIVIDUAL OR SMALL GROUP

Instructions

1. Identify a situation you have observed in a health care organization that needs to be changed.
2. What are the driving forces?

3. What are the restraining forces?
4. Develop an action plan using one of the following strategies:
   a. Power-coercive strategies
   b. Empirical-rational strategies
   c. Normative-reeducative strategies

Analysis/Discussion

1. Explain your rationale for picking the strategy you did.
2. How easy or difficult do you believe change will be in this situation?

Further Reading

Lutjens, L. R. J. & Tiffany, C. R. (1994). Evaluating planned change theories. *Nursing Management,* 25(3), 54–57.

Perlman, D. & Takacs, G. J. (1990). The 10 stages of change. *Nursing Management,* 21(4), 33–38.

Tiffany, C. R., Cheatham, A. B., Doornbos, D. et al. (1994). Planned change theory: Survey of nursing periodical literature. *Nursing Management,* 25, 54–59.

Exercise 24-2 Managing Resistance

INTRODUCTION

Resistance to change occurs because of fear of the unknown or when individuals are threatened by the change. Resistance can be beneficial or detrimental. Resistance can promote problem solving and motivate performance in an attempt to negate the need for change. Resistance can also force the change agent to clarify the need for change and to maintain a high level of interest. Conversely, resistance can redirect energies and threaten morale.

EXERCISE

TYPE OF ACTIVITY: INDIVIDUAL OR SMALL GROUP

Instructions

1. Read the following scenario.
2. Identify phrases or behaviors that indicate resistance.
3. Describe ways to handle resistance in this situation.

Scenario

Adam was recently hired as manager of the orthopedic unit. As he has become more familiar with the unit and the facility he has begun to question some of the policies and procedures. One is the policy that only orthopedic technicians trained in physical therapy can apply and change traction set-ups on the beds. As he explores with his staff the history behind this policy, he is told the nurses do not have time to "fiddle" with the traction. He also is informed that it has always been done that way. He suggests developing a task force to review the policy and make recommendations for change. He hears rumblings among the staff that "every new boss wants to do something different." Bill, his nurse specialist, informs him that they tried that once before, but the staff wasn't interested. Because the current policy has resulted in several incidents and delays in treatment, Adam does not feel he can avoid the issue. He suggests a proposal in which the unit orthopedic aides are trained in the application of traction and a sequencing of who to call in the event traction needs to be set up, adjusted, or changed. Staff members don't think it will work because the unit aides are just as busy as the nurses; besides, they contend physical therapy will never agree to the change. Several nurses suggest waiting, that "things have a way of working themselves out."

Analysis/Discussion

1. How realistic did you find this scenario?
2. How easy or difficult was it to identify resistance?
3. What did you learn about handling resistance?

Further Reading

Bolton, L. B., Aydin, C., Popolow, G. & Ramseyer, J. (1992). Ten steps for managing organization change. *Journal of Nursing Administration,* 12(6), 14–20.

Harvey, T. R. (1990). *Checklist for Change: A Pragmatic Approach to Creating and Controlling Change.* Boston: Allyn & Bacon.